Bonded Repair of Aircraft Structures

ENGINEERING APPLICATION OF FRACTURE MECHANICS
Editor-in-Chief: George C. Sih

G.C. Sih and L. Faria (eds.), Fracture mechanics methodology: Evaluation of structure components integrity. 1984. ISBN 90-247-2941-6.
E.E. Gdoutos, Problems of mixed mode crack propagation. 1984. ISBN 90-247-3055-4.
A. Carpinteri and A.R. Ingraffea (eds.), Fracture mechanisms of concrete: Material characterization and testing. 1984. ISBN 90-247-2959-9.
G.C. Sih and A. DiTommaso (eds.), Fracture mechanics of concrete: Structural application and numerical calculation. 1984. ISBN 90-247-2960-2.
A. Carpinteri, Mechanical damage and crack growth in concrete: Plastic collapse to brittle fracture. 1986. ISBN 90-247-3233-6.
J.W. Provan (ed.), Probabilistic fracture mechanics and reliability. 1987. ISBN 90-247-3334-0.
A.A. Baker and R. Jones (eds.), Bonded repair of aircraft structures. 1987. ISBN 90-247-3606-4.

Bonded Repair of Aircraft Structures

Edited by

A.A. Baker

and

R. Jones

Department of Defence
Defence Science and Technology Organisation
Aeronautical Research Laboratories
Melbourne, Victoria, Australia

1988 **MARTINUS NIJHOFF PUBLISHERS**
a member of the KLUWER ACADEMIC PUBLISHERS GROUP
DORDRECHT / BOSTON / LANCASTER

Distributors

for the United States and Canada: Kluwer Academic Publishers, P.O. Box 358, Accord Station, Hingham, MA 02018-0358, USA
for the UK and Ireland: Kluwer Academic Publishers, MTP Press Limited, Falcon House, Queen Square, Lancaster LA1 1RN, UK
for all other countries: Kluwer Academic Publishers Group, Distribution Center, P.O. Box 322, 3300 AH Dordrecht, The Netherlands

Library of Congress Cataloging in Publication Data

Bonded repair of aircraft structures.

(Engineering application of fracture mechanics ; 7)
Includes index.
1. Airframes--Maintenance and repair. 2. Metal bonding. I. Baker, A. A. (Alan A.) II. Jones, R. III. Series.
TL671.9.B63 1987 629.134'31 87-22001

ISBN 90-247-3606-4

Copyright

PRINTED IN THE NETHERLANDS

Contents

Series on engineering application of fracture mechanics

Fracture mechanics technology has received considerable attention in recent years and has advanced to the stage where it can be employed in engineering design to prevent against the brittle fracture of high-strength materials and highly constrained structures. While research continued in an attempt to extend the basic concept to the lower strength and higher toughness materials, the technology advanced rapidly to establish material specifications, design rules, quality control and inspection standards, code requirements, and regulations for safe operation. Among these are the fracture toughness testing procedures of the American Society of Testing Materials (ASTM), the American Society of Mechanical Engineers (ASME) Boiler and Pressure Vessel Codes for the design of nuclear reactor components, etc. Step-by-step fracture detection and prevention procedures are also being developed by the industry, government and university to guide and regulate the design of engineering products. This involves the interaction of individuals from the different sectors of the society that often presents a problem in communication. The transfer of new research findings to the users is now becoming a slow, tedious and costly process.

One of the practical objectives of this series on *Engineering Application of Fracture Mechanics* is to provide a vehicle for presenting the experience of real situations by those who have been involved in applying the basic knowledge of fracture mechanics in practice. It is time that the subject should be presented in a systematic way to the practising engineers as well as to the students in universities at least to all those who are likely to bear a responsibility for safe and economic design. Even though the current theory of linear elastic fracture mechanics (LEFM) is limited to brittle fracture behavior, it has already provided a remarkable improvement over the conventional methods not accounting for initial defects that are inevitably present in all materials and structures. The potential of the fracture mechanics technology, however, has not been fully recognized. There remains much to be done in constructing a quantitative theory of material damage that can reliably translate small specimen data to the design of large size structural components. The work of the physical metallurgists and the fracture mechanicians should also be brought together by reconciling the details of the material microstructure with the assumed continua of the computational methods. It is with the aim of developing a wider appreciation of the fracture mechanics technology applied to the design of engineering structures such as aircrafts, ships, bridges, pavements, pressure vessels, off-shore structures, pipelines, etc. that this series is being developed.

Undoubtedly, the successful application of any technology must rely on the soundness of the underlying basic concepts and mathematical models and how they reconcile with each other. This goal has been accomplished to a large extent by the book series on *Mechanics of Fracture* started in 1972. The seven published volumes offer a wealth of information on the effects of defects or cracks in cylindrical bars, thin and thick plates, shells, composites and solids in three dimensions. Both static and dynamic loads are considered. Each volume contains an introductory chapter that illustrates how the strain energy criterion can be used to analyze the combined influence of defect size, component geometry and size, loading, material properties, etc. The criterion is particularly effective for treating mixed mode fracture where the crack propagates in a non-self similar fashion. One of the major difficulties that continuously perplex the practitioners in fracture mechanics is the selection of an appropriate fracture criterion without which no reliable prediction of failure could be made. This requires much discernment, judgement and experience. General conclusion based on the agreement of theory and experiment for a limited number of physical phenomena should be avoided.

Looking into the future the rapid advancement of modern technology will require more sophisticated concepts in design. The micro-chips used widely in electronics and advanced composites developed for aerospace applications are just some of the more well-known examples. The more efficient use of materials in previously unexperienced environments is no doubt needed. Fracture mechanics should be extended beyond the range of LEFM. To be better understood is the entire process of material damage that includes crack initiation, slow growth and eventual termination by fast crack propagation. Material behavior characterized from the uniaxial tensile tests must be related to more complicated stress states. These difficulties should be overcome by unifying metallurgical and fracture mechanics studies, particularly in assessing the results with consistency.

This series is therefore offered to emphasize the applications of fracture mechanics technology that could be employed to assure the safe behavior of engineering products and structures. Unexpected failures may or may not be critical in themselves but they can often be annoying, time-wasting and discrediting of the technical community.

Bethlehem, Pennsylvania
1987

G.C. Sih
Editor-in-Chief

Preface

The conventional approach to through-life-support for aircraft structures can be divided into the following phases: (i) detection of defects, (ii) diagnosis of their nature and significance, (iii) forecasting future behaviour-prognosis, and (iv) prescription and implementation of remedial measures including repairs.

Considerable scientific effort has been devoted to developing the science and technology base for the first three phases. Of particular note is the development of fracture mechanics as a major analytical tool for metals, for predicting residual strength in the presence of cracks (damage tolerance) and rate of crack propagation under service loading. Intensive effort is currently being devoted to developing similar approaches for fibre composite structures, particularly to assess damage tolerance and durability in the presence of delamination damage.

Until recently there has been no major attempt to develop a science and technology base for the last phase, particularly with respect to the development of repairs. Approaches are required which will allow assessment of the type and magnitude of defects amenable to repair and the influence of the repair on the stress intensity factor (or some related parameter). Approaches are also required for the development and design of optimum repairs and for assessment of their durability.

It is the purpose of this book to discuss the emerging science for repairs based on adhesive bonding. As shown in the book, use of structural adhesive bonding allows optimised repairs to be accomplished in many situations where traditional approaches based on mechanical attachment would previously have been employed. For example, cracked metallic aircraft components are often repaired by bolted or riveted metallic reinforcements despite their relatively poor efficiency and the damage to the structure and substructure which may arise from their implementation. Adhesive bonding is of course the main approach for repairing adhesively bonded metallic components and for repairing advanced fibre composite components, particularly those with relatively thin skins.

In Chapter 1 of the book, Kelly provides a very useful general introduction to the status of bonded repairs. Reinhart, in Chapter 2, reviews the critical topic of surface-treatments for bonded repairs to metallic components and shows that, at least for aluminium alloys suitable processes are available – based on phosphoric acid anodising process. In Chapter 3, Hart-Smith provides simple guidelines for the repair of bonded metallic components.

Chapters 4 to 6 describe Australian work on the repair of fatigue or stress-corrosion cracked metallic components with boron/epoxy patches and structural film adhesives – known as 'Crack-Patching'. Design aspects are discussed in Chapter 4 by Jones which covers the finite element approach and in Chapter 5 by Rose which describes an analytical approach. A wide range of experimental aspects are covered in Chapter 6 by Baker. Chapter 6 also provides an outline of two practical applications.

Finally Chapter 7 by Trabocco, Donnellan and Williams provides an overview on the repair of damaged graphite/epoxy composites – drawing attention to the problems that can be caused by moisture absorption in the epoxy matrix of the composite.

May 1986 A.A. Baker
Aeronautical Research Laboratories R. Jones

Contributing authors

A.A. Baker
Aeronautical Research Laboratories, PO Box 4331, Melbourne, Australia

T.M. Donnellan
Code 6063, Naval Air Development Centre, Warminster, PA 18974, USA

L.J. Hart-Smith
Structural Mechanics, Mail Stop 36-90, Douglas Aircraft Co, McDonnell Douglas Corp, Long Beach California 90846, USA

R. Jones
Aeronautical Research Labs, PO Box 4331, Melbourne, Australia

L.J. Kelly
Air Force Wright Aeronautical Lab, AFWAL/FIB-LB, Wright Patterson AFB, OH 45433, USA

T.J. Reinhart
AFWAL/MLSE, Wright Patterson AFB, OH 45433, USA

L.R.F. Rose
Aeronautical Research Labs, PO Box 4331, Melbourne, Australia

R.E. Trabocco
Code 6063, Naval Air Development Centre, Warminster, PA 18974, USA

J.G. Williams
Code 6063, Naval Air Development Centre Warminster, PA 18974, USA

L.J. KELLY

1

Introductory chapter

The use of adhesive bonding as a joining method in aircraft construction is an accepted means of attaining high structural efficiency and improved fatigue life. Extensive use of this type of construction has been made in aircraft secondary structure. For example, 62% of the Boeing 747 wetted area is adhesive bonded structure and the Lockheed C5A contains 35,000 sq. ft. (3250 sq. m) of bonded structure. Some selected aircraft have employed adhesive bonding for primary structure including wing stiffener, fuselage longeron and fuselage skin panel splice areas. The most noteworthy of these is the Fokker F-27 aircraft which has had over 25 years of successful service experience (Figure 1.1).

The use of adhesive bonding in aircraft structure has expanded greatly in recent years as more and more advanced composite materials are being utilized. In general, the most efficient composite to metal splice joints have been scarf and stepped lap joints in which there is relatively little change in the load path. With this increased use of bonded assemblies comes the increased potential for processing anomalies, mistakes during fabrication, and in service damage. Repair methods are required as an option to costly scrapping of large assemblies.

Repair documents (technical orders) have been developed by each manufacturer using his own preferred processing methods and procedures. A mechanic having to repair similar damage on several different aircraft is often instructed to use different materials and methods on each. Although the environmental and structural requirements of these aircraft can differ especially between military and commercial operation, there are really only small variations in the essentials of good repair techniques.

The objective of this book is to provide a broad coverage of repair information applicable to a wide variety of bonded structure and also to describe recent work on the repair of cracked metallic structure. The information in this book should be used to supplement rather than supplant the repair technical orders for a specific aircraft. In any repair situation the technical orders must be consulted regarding such information as operating environments, damage size limits, repair proximity limits and weight and balance limits.

The materials and methods described for the repair of bonded sandwich, bonded metal sheet, and bonded composite assemblies are fairly standard. They have been selected because they have been successfully used and their service performance known to be satisfactory. They are the techniques most likely to restore damaged

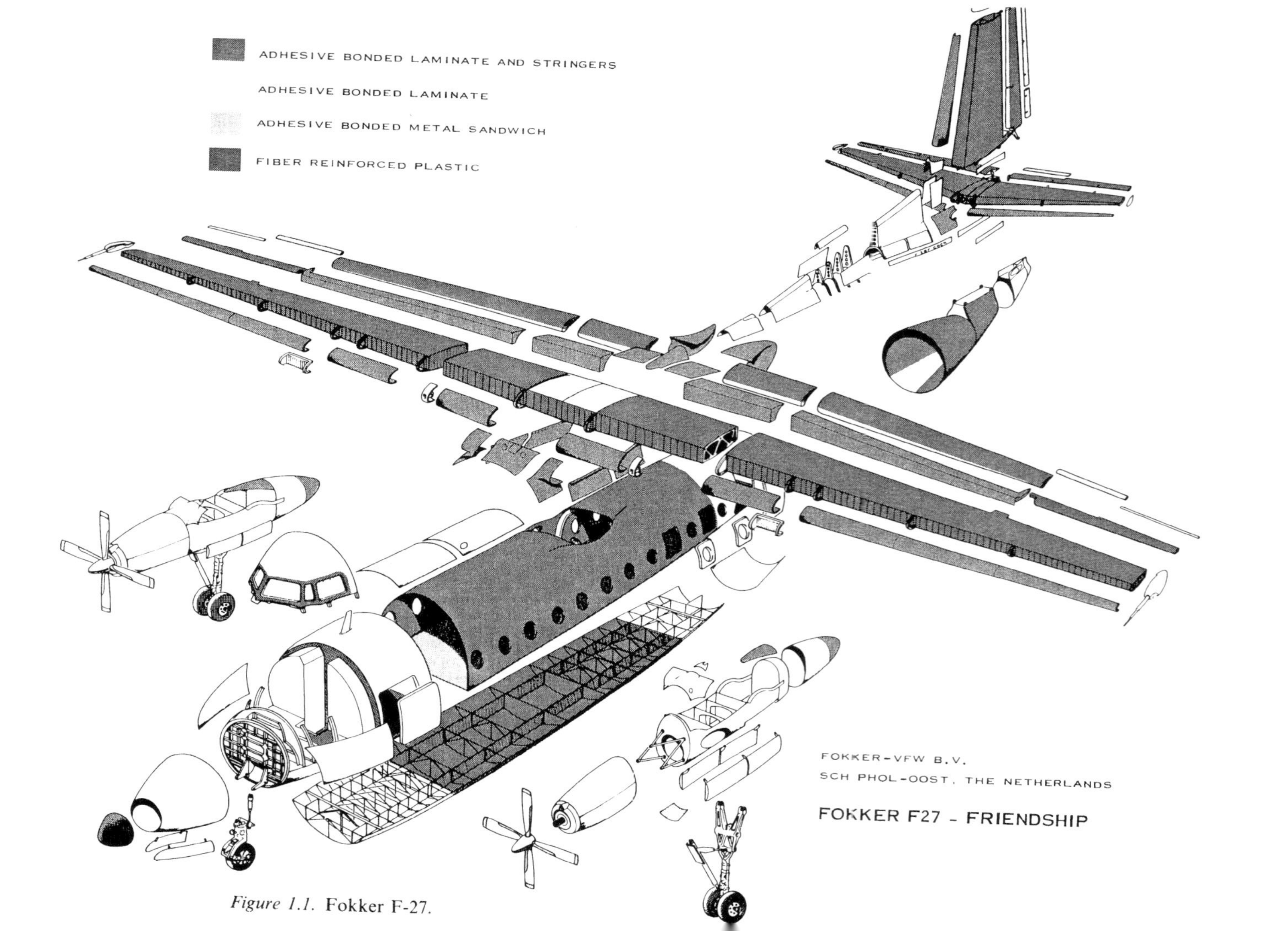

Figure 1.1. Fokker F-27.

aircraft structure to its original level of strength and durability and the technology that will provide the confidence necessary to permit bonding to expand to even wider use in primary aircraft structure. The design approach, materials and procedures described for the repair of cracked metallic components are of a more experimental nature and this is reflected in the treatment given in the relevant chapters. However, several significant practical successes have been demonstrated.

1.1 Bonded vs bolted repairs

A discussion of the relative merits of bonded and bolted repairs is really a discussion of bonded and mechanical fastened joints. The best that can be said for mechanical joints is that the joint can be subsequently disassembled and can be made in an uncontrolled environment. The machining of holes in the members to be joined to accommodate mechanical attachments obviously weakens the load carrying capability of the members and produces concentrated stresses at the bearing surfaces resulting in local stress risers. This is particularly true for composite structural assemblies which require local reinforcement such as metallic interleaving, doublers, softening strips or local ply buildup to develop acceptable bolted joint strength.

On the other hand, bonded joints should not be attempted unless stringent cleaning and processing steps can be adhered to within a controlled environment. The area where surface cleaning of the parts is done should be isolated from such operations as sanding or grinding that generate dust, oil vapors or other contaminents. Smoking or eating in the area should be prohibited. All personnel handling cleaned parts should wear white lint-free gloves. After cleaning and priming parts should be sealed in non-contaminating oil-free paper or polyethylene film. Above all the joint should be designed such that the adhesive is stressed in the direction of its maximum strength (shear) and tension or peel stresses minimized. Provided controlled processing steps are carefully followed with emphasis on those items concerned with maintenance of a properly prepared bonding surface and good prefit of part details, very efficient joints that distribute load over a large area and eliminate stress risers can be produced.

As a rule, adhesive joints prove to be more efficient for lightly loaded structure while mechanically fastened joints are more efficient for highly loaded structure. Reference [1] showed that for a boron/epoxy doubler lap joint the breakeven point was around 11,000 pounds per inch (1900 kN/m). References [2, 3] show that symmetrical stepped lap or scarf joints can raise the crossover point to 40–50,000 pounds per inch (7000–8700 kN/m).

The author of chapter 3 of this book has shown, in reference 4, that while thin to moderately thick structures have a remarkable tolerance for quite large bond imperfections the complex joints associated with bonding of thicker adherends exhibit a sensitivity to both large voids and porosity. For thin adherends even the flawed bonds are often strong enough since in a real structure in which random bond flaws are surrounded by nominal perfect bonds, any flawed bonds divert some of their share of the load to the adjacent sound bounds. The prime effect of all kinds of bondflaws is a reduction in the thickness of members that can be bonded satisfactorily. However, flaws in thick bonded structure can propogate catastrophically so mechanical fasteners are advocated as a fail safe load path. The author states further that it is best to restrict the use of adhesive bonding to those applications and designs in which there is no possibility of any local bond flaw growing during the life of the aircraft and that it is unwise to ever design or build a purely bonded joint which is weaker than the members themselves. Mechanical fasteners should be used in such a joint even if the bond is flawless. This is considered sound engineering guidance for the designer of bonded structure repairs.

The fasteners alone should be designed to carry limit load. The fasteners can then be very effective in preventing widespread unzipping by what was initially localized load redistribution around a damaged or defective bond. While the fasteners carry virtually no load, as long as the bond is intact, they enable any remaining adhesive after damage to work efficiently as well as accept load themselves to relieve the load on the bonds or resin interface in the case of composites. When bolts are used in this fashion to repair defective or damaged bonds their effectiveness follows from relieving the stress and strain concentration in the adhesive immediately adjacent to the flaw, rather than in the load they transfer themselves, in other words, the benefit from the presence of the bolts is that they permit the remaining adhesive to work more effectively.

One very effective use of bolts to achieve a structural repair of a bonded assembly is in the repair of delaminations in composites. Recent work [5–7] has shown that fracture mechanics provides by far the most promising method for assessing the severity of impact damage and the need to repair. Several researchers have observed that repairs of delaminations attempted by injecting resin into voids do not result in structural improvements. If the internal surfaces are in any way contaminated the injected resin will not bond. If the delaminations are caused by impact it has been found that multiple unconnected delaminations, are produced (Figure 1.2) that are very difficult to adequately fill by injected resin. In order to be injected the resin must have low viscosity and such resins have poor strength characteristics, especially in hot, wet conditions. For these reasons resin injection by itself should be considered a non-structural repair. On the other hand, a simple bolt clamping repair can restore the major portion of the undamaged laminate strength of impact damage composite structure that is not subject to compression dominated fatigue loading with substantial postbuckling. This latter load condition requires the combination of injection and fastening with multiple fasteners carefully spaced around the edge of the damage in line with the primary load directions.

While bolts are effective in repairing delaminations in composites, composites are equally effective, or more so, as a patch for cracked metal structure. Traditionally conventional aircraft metallic structure that encounters fatigue or stress corrosion cracking is repaired by riveting or bolting a metallic reinforcement over the damaged area. Such repairs introduce further local stress concentrations at the additional fastener holes which in turn can result in increased fatigue cracking. The application of high modulus composites in conjunction with high performance

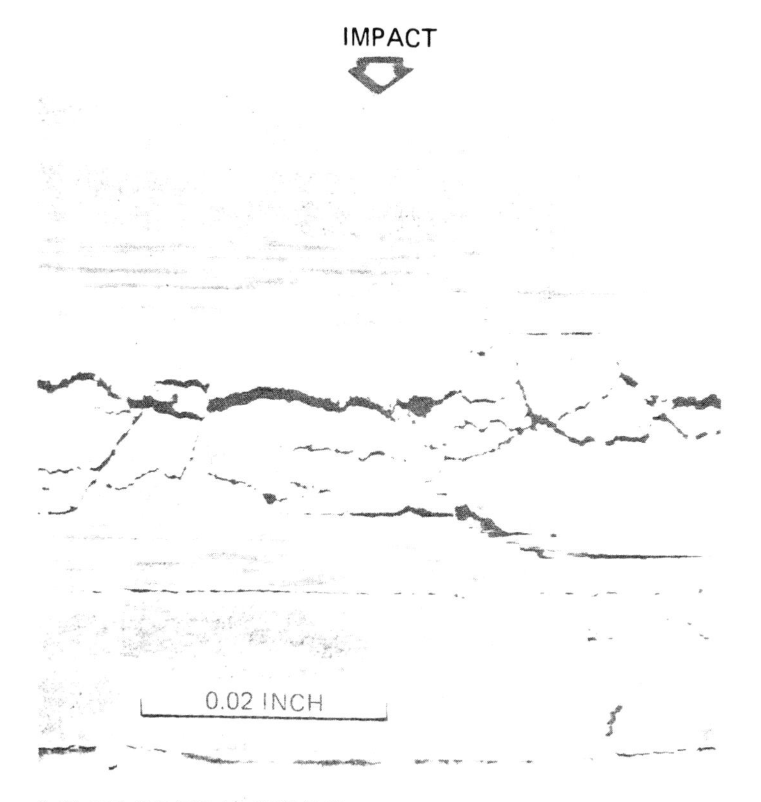

8 PLIES GR/EP $(\pm 45/0/90)_S$

BLUNT IMPACTOR AT CENTER OF 5 INCH SQUARE AREA
TOTAL ABSORBED ENERGY = 1.24 FT-LB
(INCIPIENT DAMAGE INDICATED AT 0.82 FT-LB)
DAMAGE NOT VISIBLE ON IMPACTED SURFACE
SLIGHT MATRIX CRACK ON BACK FACE

Figure 1.2. Photomicrograph of impact damaged laminate.

adhesives provides a method of repair that eliminates these stress concentrations. The technique, referred to as crack patching, has been pioneered by the Aeronautical Research Laboratories Australia and is discussed in detail in chapters 4, 5 and 6. The design of a composite patch is based on the principles of fracture mechanics. Two design approaches have been developed, both involve the use of fracture mechanics. The finite element approach for the detailed design of complex repairs is discussed in chapter 4, whilst the development of analytical formulae for the initial design process is discussed in chapters 5 and 6.

Although this repair procedure was initially developed for thin metallic components analysis and laboratory test results have shown that it is also applicable to components up to 1/2 inch (i.e. 12.7 mm) thick (see chapter 4).

Conceptually, composite patches reinforced with high modulus fibers can be expected to restrain opening deformations of cracks occurring in metal components to which the patches are bonded, thereby reducing crack tip intensities and hopefully achieving significant reductions in crack growth rates. The application of both boron and carbon epoxy composite bonded patches to cracked metal structure has indeed proved very successful in reducing crack growth rates and extending the useful life of damaged aircraft metallic structure. There are obvious advantages and disadvantages for either material but both are acceptable from a consideration of strength and stiffness and both have been successfully utilized. Boron prepreg is more difficult to cut, less flexible when required to conform to curved surfaces, but on the other hand, being stiffer will pick up more load and has a higher thermal expansion coefficient (which reduces the severity of the residual thermal stresses). Carbon prepreg is cheaper and more readily conforms to curved surfaces and, although it has potential galvanic corrosion problems with aluminium, it can be easily isolated from aluminium with the application of a glass fabric between the adhesive and the first ply of prepreg.

Proper design of the patch only requires that the patch absorb an appreciable fraction of load imposed on the structure in the vicinity of the patch and that the patch does not debond from the structure under extensive load cycling. For a fixed material type, thickness and loading condition one can vary patch width thickness and ply orientation to achieve a highly efficient patch design. However, a few standard patch thicknesses and ply orientations have been found to be effective in repairing a wide range of crack lengths indifferent alloys and structure thicknesses by varying only the patch size. Such patches have been employed on surface scratches, through the thickness cracks, cracks and holes and cracks at lands and root radii.

No matter whether boron or carbon composite patches are used the following general design guidelines apply. The ply orientation should be selected to have the maximum practical number of 0° plies in the direction perpendicular to the crack. A small number of plies could also, if desired, be oriented at 90° and ±45° to these plies to prevent cracking under structure biaxial and shear loads. To minimize shear stress concentrations in the adhesive as the load is transferred to the patch, ends of the patch should taper in the direction of load by dropping off plies. The number of 0° plies (running normal to the crack) which end at any one step should be minimized. In addition serrating the ends of the 0° plies nearest the metal surface can help to reduce the peak shear stress. Serrating can be easily done by cutting across the end of the ply with ordinary dressmaker's V-notch pinking shears. Either a 0° or a pair of ±45° plies should be next to the metal surface. The outer most plies should be a pair of 45° plies since they are less strength critical if damaged. Extending the outermost ply over the ends of all other plies provides sealing and reduces any tendency for individual ply peeling.

The bulk of the composite patching data available to date [6–10] is for 2024-T3 and 7075-T6 aluminium alloys. These results (as much as a factor of 10 improvement in fatigue life) are not directly applicable to alloys with significantly different crack growth characteristics. However, equally dramatic fatigue life improvements could be expected when composite patches are applied to alloys whose crack growth rate (da/dn vs ΔK) is slower and whose fracture toughness (K_c) is higher.

In addition, even though the results would be overly conservative, the patches developed for through cracks in chapter 6 could be used for surface scratches by assuming the surface scratch to be a through crack of equal length.

Prepregs and adhesives which require cure temperature above 250°F (121°C) have the obvious disadvantage of potential thermal damage to the aluminium structure and result in thermal mismatch problems in repairing metals in general because of differences in the coefficients of thermal expansion. Residual thermal stresses which develop after high temperature curing or bonding will cause tensile residual stresses in the metal and increase the stress intensity factors adversely affecting crack growth rate. Figure 1.3, from [8], shows a typical variation of stress intensity factor with half crack length 'a' for 7075-T6 aluminium adhesively bonded to a 12 ply $(\pm 45/0_2/90/0)_s$ patch for three values of temperature difference ΔT. However, as noted in chapter 6, in a typical aircraft structure constraint against local thermal expansion will reduce the effective expansion coefficient of the metal and thus significantly reduce residual stress compared to the maximum values predicted.

Room temperature curing adhesives, on the other hand, do not have adequate strength and durability at elevated temperature service. Adhesives that cure at temperatures above 250°F (121°C) tend to be more brittle and have lower peel strength. The adhesives that cure in the 200–250°F (93–121°C) range have good strength and stiffness and adequate moisture resistence so that they provide the best compromise for composite patching.

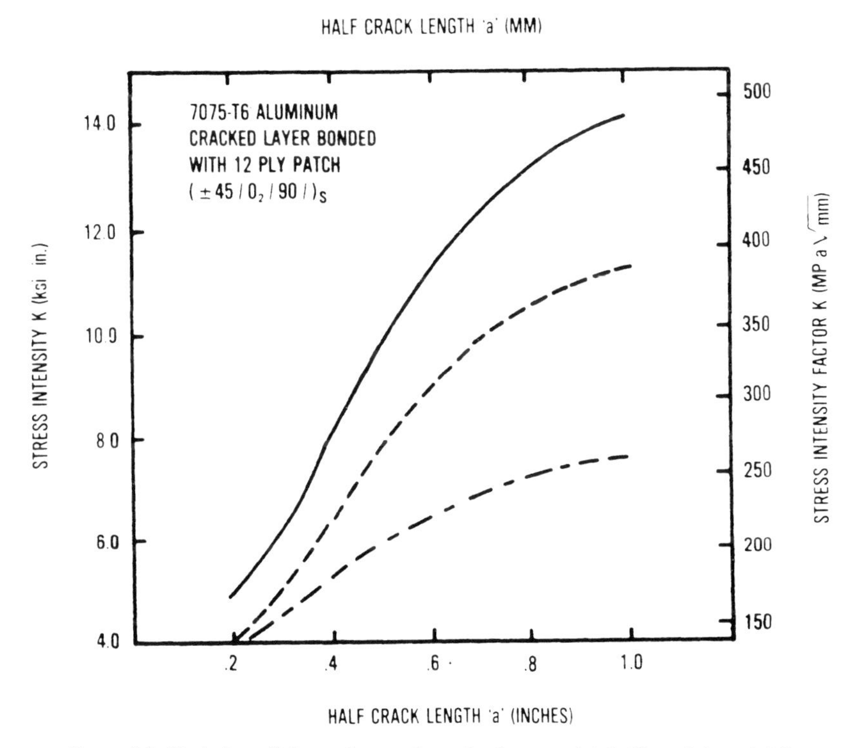

Figure 1.3. Variation of thermal stress intensity factors with half crack length 'a'.

1.2 Combined bonded/bolted repairs

In [11], see Figure 1.4, boron epoxy laminates bonded and bolted to 7075-T6 aluminium in a double overlap configuration produced better results than similar 'bolted-only' or 'bonded-only' joints.

Based on a nominal bond area of 3 square inches (1-inch (25.4 mm) wide × 1.5-inch (38.1 mm) overlap), the average bond shear strength was 4,300 psi (29.6 MPa) for the 'bolted-and-bonded' specimens, compared with an interpolated value of 3,300 psi (23 MPa) for a bonded only configuration. A similar bolted only configuration produced a joint strength only 31 per cent of the bonded configuration and only 24 per cent of the bolted-and-bonded configuration. These tests used a ductile type adhesive (Shell 951). It was said that the presence of the bolt enhanced the performance of the bond, and vice versa.

The reader is cautioned that this in not always the case. Indeed reference [12] shows that the addition of fasteners to an unflawed adhesive bonded joint does not always increase the joint strength but may actually decrease it because of a change in failure mode or location. A combined bonded/bolted joint analysis performed in [12] shows that the combination is generally no better than a nominally perfect adhesive bonded joint alone, because the adhesive is typically so much stiffer than the fasteners. The combination, however, has substantial benefits in the context of repair of improperly bonded structure, of the in-service repair of damaged struc-

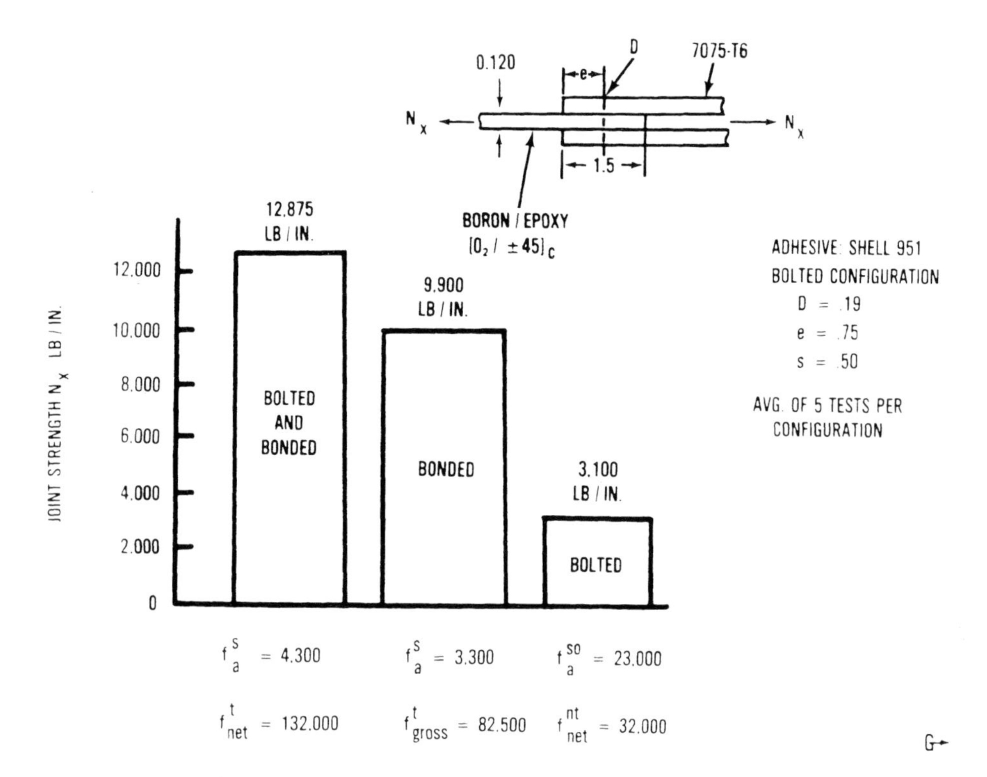

Figure 1.4. Relative strength of bolted-and-bonded joint vs bolted-only or bonded-only joints.

ture and of damage confinement in thick fibrous composites or bonded laminated metal structure.

1.3 Adhesives

The choice of adhesives for any bonded repair should be based on the strength requirements over the expected service temperature range and the type of equipment available for accomplishing bonding. There are a multitude of adhesives that provide different ranges of adhesive bonding shear and peel strengths at various service temperatures. The following is a discussion of the advantages and disadvantages of each type of adhesive and the necessary considerations in its selection. The discussion is followed by a tabular listing of currently available commercial adhesives. Both are obtained from reference [13].

Adhesives for structural bonding can be categorized into three main physical forms in which they are used: (1) films, (2) pastes, and (3) foams. Although films are easier to handle and provide a more uniform bondline thickness than paste adhesives, repair part inaccessibility or a lack of refrigerated storage equipment sometimes necessitates use of paste adhesive. Foam adhesives are used for stabilizing and splicing pieces of honeycomb core.

Film adhesives are made by blending high molecular weight polymers with curing agents, fillers, and other compounding ingredients and then casting them into thin films. In a supported film, the mixture is cast into a mesh carrier cloth. The carrier cloths in supported films are available in either random mat, knit woven, or square patterns in dacron, nylon, or glass fabrics. The carrier cloth in supported adhesive films provides better handleability than in unsupported films and helps reduce galvanic corrosion. Film adhesives are available in thicknesses ranging from five mils to 20 mils or in weights ranging from 0.03 to 0.20 lb/ft^2 (i.e. 0.15 Kg/m^2 to 1.0 Kg/m^2). It is recommended in repair procedures that a 51- to 20-mil thickness adhesive be used for composite- or metal-to-core applications and a 6- to 10-mil adhesive be used for metal-to-metal and composite-to-composite bonds.

Film adhesives offer various advantages over paste adhesives. The major advantages of film adhesives are that they are easier to apply and do not require mixing equipment. Film adhesives have more uniform viscosity and composition and provide more bondline thickness uniformity in a joint than do paste adhesives. The major disadvantages of films is that refrigeration is required for storage. In addition, films are more expensive than pastes and require heat and pressure to achieve satisfactory bonds.

Paste adhesives are formulated for application by spatula or other spreading equipment. They are available in one or two component adhesive forms that can be room temperature or heat cured. The components in a two-part paste consist of a liquid epoxy resin and a cross-linking agent. Two-component paste adhesives have a long shelf-life and do not have to be refrigerated. However, they have to be mixed before application, which increases labor costs and introduces possible

Table 1. Adhesive pastes.

Service temperature	Product name	Supplier	Cure cycle	Form	Storage	Remarks
– 67°F or 180°F	EA-9309	Dexter Corp/ Hysol	3 days @ R.T. or 1 hr @ 150°F	2-Part	12 mo @ R.T.	At R.T., 24 hours is sufficient for quick service
– 67°F to 200°F	EA-9320	Dexter Corp/ Hysol	5 days @ R.T. or 15 min @ 250°F	2-Part	12 mo @ R.T.	At R.T. 24 hours is sufficient for quick service
– 67°F to 200°F	EC-1386	3M Co.	60 min @ 350°F	1-Part	40°F or Below	
– 67°F to 250°F	EC-2214	3M Co.	40 min @ 250°F	1-Part	40°F or Below	Available in 3 forms
– 67°F to 300°F	EA-9432	Dexter Corp/ Hysol	60 min @ 250°F or 30 min @ 300°F	1-Part	6 mo @ R.T. or 12 mo @ 40°F	
– 67°F to 300°F	EA-9414.1	Dexter Corp/ Hysol	60.min @ 250°F or 20 min @ 350°F	1-Part	3 to 4 mo @ R.T. or 12 mo @ 40°F or below	
– 67°F to 350°F	EC-2258	3M Co.	60 min @ 350°F	1-Part	3 mo @ R.T. or 12 mo @ 40°F or below	
– 67°F to 350°F	EA-934A/B	Dexter Corp/ Hysol	6 days @ R.T. or 1 hr @ 200°F	2-Part	Part A 3 mo @ R.T. Part B 24 mo @ R.T.	
– 67°F to 400°F	EA-9321	Dexter Corp/ Hysol	5 days @ R.T. or 1 hr @ 180°F	2-Part	2 to 3 mo @ R.T. or 24 mo @ 40°F or below	At R.T., 24 hours is sufficient for quick service

Temperature conversions for Tables 1, 2 and 3. – 67°F ≡ – 55°C; 200°F ≡ 93°C; 250°F ≡ 121°C; 180°F ≡ 82°C; 250°F ≡ 121°C; 300°F ≡ 149°C; 400°F ≡ 204°C

Table 2. Adhesive foams.

Service temperature	Product name	Supplier	Cure cycle	Thickness and cured density	Expansion ratio	Storage
−67°F to 180°F	FM-39	American Cyanamid	1 hr 250°F	0.050, 0.100″ 20 to 40 lb/ft.^{3}	1.7	6 mo 0°F
−67°F to 180°F	EA-9818	Dexter Corp Hysol	1 hr 250°F	0.050, 0.100″ 42 lb/ft^3	1.7	6 mo 0°F 1 mo 32°F 7 days 75°F 5 days 90°F
−67°F to 350°F	FM-40	American Cyanamid	1 hr 350°F	0.025, 0.050, 0.100″ 20 to 40 lb/ft^3	1.7 to 3.0	6 mo 0°F 5 days 90°F
−67°F to 350°F	AF-3002	3M	1 hr 350°F	0.050″ 35 lb/ft^3	1.6	6 mo 0°F 7 days 75°F
−67°F to 350°F	FM-404	American Cyanamid	1 hr 350°F	0.200″ 12 to 35 lb/ft^3	4.0 to 5.0	6 mo 0°F

human error with incomplete mixing or improper weighing. Currently available paste adhesives have lower strength properties than film adhesives, especially for elevated temperatures service.

Foam adhesives for structural bonding are generally formulated with epoxy resin systems that contain a foaming agent so that they will expand during a cure cycle. They are used in honeycomb repair to fill gaps in splice areas or between edge members and honeycomb core. In addition, they are used to fill voids and eliminate moisture paths through splice areas.

Foam adhesives can be obtained in either paste or tape (film) forms. They have expansion ratios ranging from 1.3 to 5.0 and densities ranging from 12 to 45 lb/ft^3 (192 to 720 kg/m^3) after cure. Tape thicknesses range from 0.025 to 0.200 inches (i.e. 0.35 mm–0.508 mm). The film thickness selection will depend on the size of the cells in the core or gaps in the splice area. Generally a 0.05 inch (1.27 mm) thick adhesive foam tape is used for 1/8-inch (3.175 mm) cell core, whereas a 0.100 inch-thick (2.54 mm) adhesive foam tape is usedj for 3/16- (4.76 mm) and 1/4-inch (6.35 mm) cell sizes.

Before proceeding to the detailed bonded repair techniques discussed in subsequent chapters the reader should not only be familiar with available adhesives but also adhesive qualification testing, adherend surface preparation and environmental influences on bonded repairs.

1.4 Adhesive testing

Standard adhesive receiving acceptance tests such as the following are commonly employed:

Table 3. Adhesive films.

Service temp.	Product name	Supplier	Cure cycle	Film weights	Storage	Remarks
−67°F to 180°F	FM-73	American Cyanamid	1 hr 250°F	0.03, 0.045, 0.060, 0.085 psf	6 mo 0°F	
−67°F to 180°F	EA-9628	Dexter-Hysol	90 min 200°F or 1 hr 250°F	0.06, 0.085 psf	6 mo 0°F	
−67°F to 180°F	AF-163-2	3M	1 hr 250°F	0.06, 0.085 psf	6 mo 0°F	
−67°F to 300°F	FM-300	American Cyanamid	1 hr 350°F	0.85 psi	6 mo 0°F	
−67°F to 300°F	AF-147	3M	90 min 295°F or 1 hr 350°F	0.075 psf	6 mo 0°F	High peel version of AF-143
−67°F to 350°F	FM-300	American Cyanamid	1 hr 350°F	0.08, 0.10 psf	6 mo 0°F	
−67°F to 350°F	AF-143	3M	1 hr 350°F	0.075, 0.10 psf	6 mo 0°F	
−67°F to 350°F	Metlbond 1515	Narmco	1 hr 350°F	0.030, 0.060, 0.080 psf	6 mo 0°F	
−67°F to 400°F	FM-400	American Cyanamid	1 hr 350°F	0.075, 0.10 psf	6 mo 0°F	Aluminum Filler
−67°F to 400°F	Metlbond 328	Narmco	90 min 325°F or 15 min 365°F	0.095 psf	6 mo 0°F	
−67°F to 420°F	EA-9649	Dexter Corp Hysol	1 hr 350°F	0.030, 0.075, 0.10 psf	6 mo 0°F	Aluminum Filler
−67°F to 450°F	Metlbond 329	Narmco	2 hr 300°F or 1 hr 350°F or 30 min 365°F	0.095 psf	6 mo 0°F	

Test	Comment
1. Lap Shear	Should be performed at RT and the highest used temperature
2. Metal Peel	Should be performed at RT or −65°F (−54°C)
3. Honeycomb Peel	Should be performed at RT or −65°F (−54°C)
4. Film Weight	

These tests help to assure uniformity in the adhesive product but provide no direct data useful in establishing design allowables for a bonded repair. The standard lap shear test failure is more the result of joint deflection and induced peel stresses than the shear strength of the adhesive. It is important for establishing design allowables to select test coupons which generate data as closely representative of how the adhesive will be loaded in the structurally configured repair joint as possible. Practical structural joints (long overlap) react differently to load cycle rates than short overlap test coupons. In addition, because of the time dependent properties of adhesives, test load cycles must be designed to represent the real life spectrum of the structure to be repaired as nearly as possible or practical.

In the program discussed in reference [14] emphasis was placed on slow cycle testing because high frequency (30 Hz) testing of adhesive bonds led to misleading results. The adhesive would never fail because the adhesive had no time to creep. This program demonstrated that adhesives are significantly visoelastic and their response to various loads to which they are subjected are influenced by the duration of the load cycles. Short overlap thick adherend coupons tested at high frequency typical of metal fatigue tests (30 cycles per second) showed no indication of the slightest damage after 10^7 cycles. Yet the very same specimens, when tested more slowly (one or two cycles per hour) to approximate the real time service exposure in a pressurized fuselage would fail after the application of only a few hundred load cycles of the same intensity. Slow cycles cause creep accumulation and rupture (short fatigue life) in short overlaps; however, in long overlap joints creep is stabilized and long fatigue life obtained.

1.5 Surface Preparation

The shear strength achieved in any bonded repair is very much dependent on the surface preparation of the adherends. Since metal adherend surface preparation is covered in more detail in chapter 3 and composite adherend surface preparation in chapter 7 only general guidelines will be covered here.

Surface pretreatment of metal adherends to provide a clean surface with a high surface energy is an extremely important step in achieving a durable long term adhesive bond. Metal surface preparation includes removing organic contaminates and subsequently etching the surface with chemicals or placing an active oxide on the surface by anodizing. Aluminium preparations include the Forest Product Laboratory (FPL) etch (aqueous solium dichromate sulfuric acid solution); or Fokker VFW (chromic-sulphuric acid etch and chromic acid anodize) or preferably the Phosphoric Acid Anodize (PAA) Boeing Process Specification BAC5555 pre-treatment which offers improved environmental stability. Titanium adherends can be acid cleaned with the Pasajell 107 process by dipping the part in a tank solution or by applying a paste to the titanium bonding surface. Reference [15] points out that the most durable titanium surface bonds are achieved with adherends having high surface roughness and the ability to mechanically interlock with the adhesive. Good results were achieved through chromic acid anodize pre-treatment. Reference [16] indicates that simple cleaning and treatment with a 0.1 per cent aqueous solution of gamma-aminopropylepoxysilane (APS) can be a very effective surface preparation for titanium that can be readily accomplished in the field. Composite adherend surface preparation consists of wiping the faying surface with a cheesecloth containing a solvent, lightly sanding with silicon carbide abrasive paper being careful to not break the fibres and wiping repeatedly with cheesecloth containing solvent or water until no further indication of residue is visible on the cheesecloth. The laminate should stand at least 30 minutes to ensure complete evaporation of the solvent before bonding. The authors of chapter 7 recommend the final wiping be done with distilled water rather than a solvent.

A cleaned adherend should never be touched with bare hands and should be bonded as soon after the cleaning operation as possible. Preferably, a cleaned metal part should be coated with an adhesive primer for maximum protection. After the adhesive primer is applied and cured, a metal adherend can be stored for a month or more in a protective wrapping (wax free paper).

1.6 Environmental Behaviour

The author of reference [4] points out that it is significant that, in the service record of adhesively bonded metallic structure, it has been inadequate surface preparation or the use of environmentally sensitive adhesives without corrosion-inhibiting primers that has led to problems, not the incidence of various flaws in the bonds. Indeed most typically lightly loaded thin bonded structure has very considerable tolerance for all bond flaws and does not need structural repairs unless the bond defects are significantly large enough to alter the distribution of the load transfer causing a shift to a critical location. It is suggested that all that is needed is to seal the edges against ingress of moisture and corrosion for any attempt to perform a structural repair usually ensures only a decrease in life of the structure by breaking the adherend surface protection.

The Boeing crack extension test, Figure 1.5, has been found to be very discriminating relative to surface preparation quality. This test consists of two 1/8 inch (3.175 mm) thick metal plates 6 in. × 6 in. (i.e. 152 mm × 152 mm) bonded together and then cut into 1 inch (25.4 mm) wide test pieces. A strip approximately 0.75 inch (19.05 mm) wide on one edge of the panel is left unbonded. A wedge is carefully driven (using several light taps) between the plates to precrack the unbonded end of the specimens and the initial crack length noted and measured after exposure to selected environmental conditions (typically 120°F ± 5°F (i.e. 49°C ± 2.8°C) and 95% relative humidity for 1 hour, 4 hours, 72 hours, 14 days or 30 days).

Failure modes may be determined by splitting the specimen open at the com-

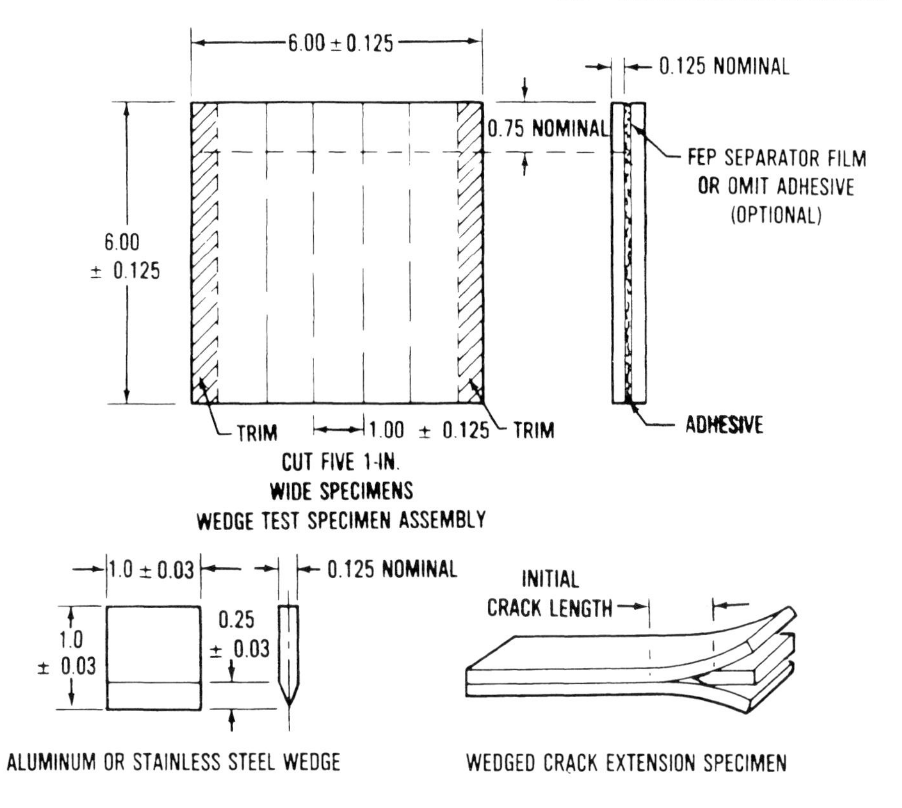

Figure 1.5. Wedge test specimen configuration.

pletion of the final crack growth measurement and examining the surface. The percentage of adhesive versus cohesive failure in the exposed crack growth region should be recorded. If specimens show less than 100% cohesive failure when peeled apart, a further evaluation should be conducted. Both the initial crack length and the rate of crack growth indicate whether the surface preparation is good or bad. Reference [17] provides the following guidelines when the wedge test is used with 250°F (121°C) curing epoxies.

1 hr crack growth, in.*	Rating
0.0 to 0.10	Very good
0.11 to 0.25	Good
0.26 to above	Marginal to Non Acceptable

*For processing control, the average of 10 specimens should not exceed 0.25 in. (6.35 mm), and no individual value should exceed 0.75 in. (19.05 mm) to be acceptable.

Reference [18] provides another interpretation of the wedge test based on user service experience and consultation with Boeing that is stated as follows:

Permanent primary structure repair. Crack growth should not be more than 1/2 in. (12.7 mm) after immersion in water at 68°F (20°C) for one hour.

Permanent secondary structure repair. Crack growth should not be more than 1-1/2 in. (38.1 mm) after immersion in water at 68°F (20°C) for one hour.

Temporary primary structure repair. Crack growth should not be more than 1-1/2 in. (38.1 mm) after immersion in water at 68°F (20°C) for one hour.

Temporary secondary structure repair. No wedge test requirement provided that the lap shear strength is considered adequate with the surface preparation used.

One particularly detrimental environment that can prevent achievement of a good quality bond is high humidity. Not only is a clean dry area (5%RH) preferable but the moisture content of the structure being repaired must be low. Moisture absorbed in a complete laminate or moisture entrapped in a bonded honeycomb assembly can be very detrimental to the integrity of a bonded repair. Carbon and boron epoxy composites can absorb as much as 1.2% of their weight in moisture during service in a humid environment. Porous bond lines, Figure 1.6, have resulted when carbon epoxy patches were bonded to carbon epoxy laminates that were not totally dried.

Moisture can have detrimental effects on bonded repairs in the following ways:

(1) Local delamination or blistering in parent laminates.

(2) Reduced strength of the repair patch and repair bond line resulting from porosity.

(3) Expanding moisture in honeycomb cells has created sufficient pressure to separate the cover skin from the core.

(4) Reduced effectiveness of ultrasonic inspection due to strong signal attenuation making it difficult to verify bond line integrity.

Prebond drying (a minimum of 48 hours at 170°F–200°F, i.e. 77–93°C) slow heat up rates, reduced cure temperatures and selection of adhesives less sensitive to moisture can minimize the likelihood of occurrence of the above problems. Drying the parent laminate to an average moisture content of less than 0.5 per cent is

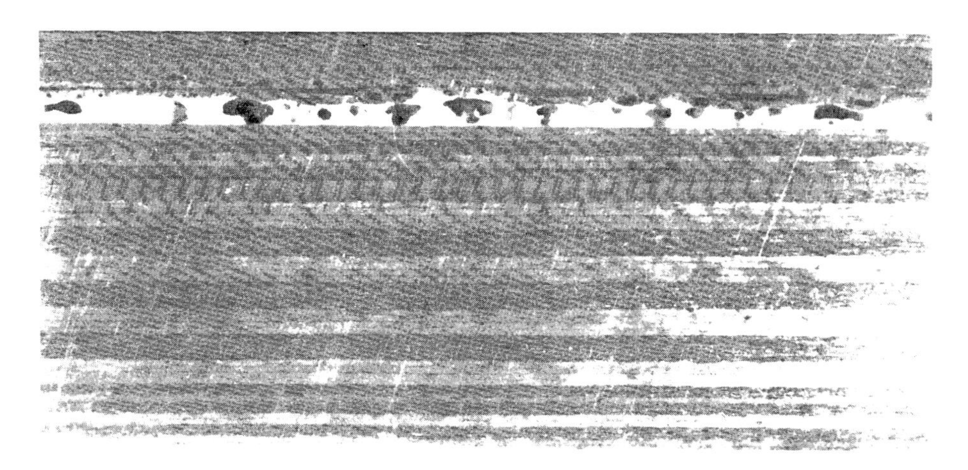

Figure 1.6. Porosity in AF-147 bondline on 50-ply wet laminate.

recommended. The use of a vacuum bag over the repair area may increase the drying rate.

1.7 Summary

The purpose of this book is to provide the repair designer guidance and a basic understanding of the principles of bonded repairs that are applicable to both secondary and primary adhesively bonded structure employed on both military and commercial aircraft. The techniques and procedures described are based on experience, analysis and demonstrated test results but to assure required structural integrity it is recommended that any major bonded repair being implemented be supported by an elemental test program to update the specific concept, materials, load transfer and process procedures.

Adhesive bonding is stiffer than riveting and fails at a lower relative displacement between the members being joined. Consequently bonded repairs are less forgiving with respect to poor detailing, load redistribution and damage tolerance than conventional riveted repairs. This demands a great level of understanding and proficiency in the design of bonded repairs and dedication to sound processing procedures required to accomplish good bonding. The reader is expected to have a familiarity with the basic principles of load transfer through bondlines.

The concept of extending the life of cracked metal structure with composite patches is very effective. The selection of adhesive for such bonded repairs or any other bonded repair should be based on modest shear strength with high peel and excellent resistance to moisture absorption. The latter requirement is the most important, for the most severe limitation to the use of structural adhesives is the susceptibility of the bondline to attack by moisture. The effect of the moisture being weakening by hydration of the oxide on the metal adherent and loss of adhesive shear strength with temperature.

Bonding *and* selective riveting should be considered a viable option along with pure bonding. Rivets can provide resistance to peel forces and a fail safe load path for bonded repairs. The surface-treatment of the adherends in the bonded repair is as important a part of the bonding system as the adhesive, the primer or the adherends themselves. Even good adhesion at nominal temperatures and environments may not be adequate in warm, humid exposure. Testing in these environments is the only method for assurance of a proper surface treatment.

References

[1] AFFDL-TR-69-43, Vol. I and II, Investigation of Joints and Cutouts in Advanced Fibrous Composites for Aircraft Structures – Joint and Attachment Investigation, (1969).

[2] AFML-TR-70-261, Advanced Development of Boron Composite Wing Structural Components, (1970).

[3] AFML-TR-70-231, Advanced Composite Wing Structures – Vol. I Engineering, (1970).

[4] Hart-Smith, L.J., AFWAL-TR-82-4172, (1982).

[5] Jones, R., Broughton, W., Mousley, R.F. and Potter, R.T., *Composite Structures*, 3, 2, pp. 167–187 (1985).

[6] Baker, A.A., Jones, R. and Callinan, R.J., *Composite Structures*, 4, 1, pp. 15–45 (1985).
[7] Gillespie, J.W. and Pipes, R.B., *Composite Structures*, 2, pp. 49–69 (1984).
[8] Ratwani, M.M., NADC-80161-60, Development of Composite Patches to Repair Complex Cracked Metallic Structures Vol. I, (1982).
[9] Baker, A.A., Proceedings of International Workshop on Defense Applications of Advanced Repair Technology for Metal and Composite Structures, Repair of Cracked or Defective Metallic Aircraft Components with Advanced Fibre Composites, (1981).
[10] Cannon, LT R. and Sandow, F., Southeastern Conference of Theoretical & Applied Mechanics, (1984).
[11] DOD/NASA Advanced Composites Design Guide Vol. IA Design, Chapter 1.3 Joints, (1983).
[12] Hart-Smith, L.J., AFWAL-TR-81-3154, (1982).
[13] Advanced Composite Repair Guide, Prepared by Northrop Corporation for United States Air Force, Air Force Wright Aeronautical Laboratories, (1982).
[14] AFFDL-TR-77-107, Primary Adhesively Bonded Structure Technology (PABST) General Material Property Data, (1978).
[15] Ditchel, B.M. and Breen, K.R., 12th National SAMPE Technical Conference, (1980).
[16] Latham, Y. and Steiniger, D., NASC Workshop on Shipboard Repair of Composite Structure and Material, Silane Promotion for Titanium Bonding, (1980).
[17] AFML-TR-76-201/AFFDL-TR-76-13, Adhesive Bonded Aerospace Structures Standardized Repair Handbook, (1976).
[18] Chan, S.Y.T. and Armstrong, K.B., Proceedings of International Workshop on Defense Applications of Advanced Repair Technology for Metal and Composite Structures, (1981).

THEODORE J. REINHART

2

Surface treatments for bonded repair of metals

2.1 Introduction

Surface preparation of the metal adherend is the keystone upon which the structural adhesive bond is formed. Extensive field service failure analysis experience in the structural adhesive bonding of both aluminum and titanium alloys has repeatedly demonstrated that bond durability and longevity are dependent upon the stability and bondability of the adherend surface [1, 2]. Combinations of heat moisture and stress have been shown to be particularly effective in sorting out or discrimination between various surface preparations used in the prebond conditioning or treatment of both aluminum and titanium alloys.

Extensive amounts of research and development activity have been directed towards the development of prebond surface treatments for both aluminum and titanium structural alloys [3]. The patent and technical literature abound with chemical treatments for these alloys, all of which are designed to provide the best bonding surface possible.

Early work in the development of surface preparations, such as, the Forest Products Laboratory (FPL) etch for aluminum alloys was hampered by the inability to accurately characterize the chemical nature of the oxide produced by the treatment as well as by the lack of a discriminating test technique [4]. This situation necessitated the fabrication, environmental aging and testing of large numbers of adhesive bonded test specimens. When one considers the number of variables involved in such a treatment it was indeed a remarkable achievement that the researchers came so close to optimizing the process.

While it should be understood that the satisfactory performance of the bonded joint is the desired object, new technology has made it possible [5] to characterize in detail the chemical and physical characteristics of surface oxide layers on metals. Instrumental techniques such as the scanning and transmission electron microscopes (SEM and STEM) and surface analysis techniques such as Auger and ion microprobe, etc; allow an intimate knowledge of the influence of surface preparation variables on the oxide produced to be gained. Thus with the availability of these new and powerful analytical tools significant new understanding of the required chemical and physical characteristics of metal pre bond surface can be expected [6]. It can also be expected that these instrumental analysis techniques will play a significant role in the control of the reliability and reproducibility of metal bonded structures. Past surface treatment evaluation techniques were limited to the stressed exposure of lap shear specimens or perhaps hot/wet peel testing. These tests, however, did not readily discriminate between surface treatments of varying durability. It was not until the recent development of the wedge opening (double cantilever beam) test by the late Bert Bethune *et al.* at the Boeing Corp. that a test for the discrimination of surface preparations became available.

In general high performance structural adhesive bonding requires that great care be exercised throughout the bonding process to ensure that quality is built into the bonded product [7]. Chemical composition control of the polymeric adhesives, strict control of surface preparation materials and process parameters as well as of the adhesive layup, part fit up, tooling and the curing process are all required to produce flight worthy structural assemblies. This, of course, is in contrast to the mechanical joining of metallic components where the general level of technology required to obtain satisfactory performance is much lower. Given this situation and the inherent advantages of adhesive bonding compared to mechanical attachment we then always strive for surface preparations that provide optimum adhesion and maximum environmental resistance at least where expensive aerospace equipment is concerned. Presently used surface preparation procedures involve several common basic steps or processes. These are as follows [8]:

1. descale
2. degrease – rinse
3. deoxidize – rinse
4. chemical etch/anodize – rinse and dry
5. prime

Structural aerospace grade aluminum and titanium alloys rarely if ever require descaling treatments, though a mild abrasive treatment of the titanium materials prior to chemical treatment is common.

Degreasing procedures consist of both hot water detergent and condensing solvent vapor cleaning treatments to remove soluable contaminants. Rinsing steps will not be dwelled upon here but it should be understood that proper rinsing procedures, normally spray rinsing using treated water, are critical to the entire surface preparation operation.

After surface scale and soluble contaminants have been removed from the metallic surface we are ready to consider a step to pre-condition the surface before the actual formation of our bonding surface. This step is usually called deoxidation and involves the use of buffered alkaline or acid solutions and perhaps elevated temperatures. The intention here is to remove oxides formed during fabrication of the metal adherend and to expose base or fresh metal surface for the next step in the process.

The formation of a stable oxide on the surface of metals to be bonded has been the subject of intense and continuing research. Work on aluminum alloys in recent times has seemingly tapered off with the introduction of the Boeing phosphoric acid anodize (PAA) by the late Bert Bethune and his coworkers [9]. Though the data showing that the PAA treatment in combination with a corrosion inhibited primer is superior in almost all aspects compared to any other method of surface preparation for aluminum alloys it is still not, for various reasons, universally accepted by the aerospace industry.

The situation concerning titanium alloys is not quite so clear cut as it is with the aluminum materials [9]. For moderate temperatures, approximately 350°F (177°C) short times, about 250 hours or less at 400°F (204°F) to 600°F (315°C) in air, there are a number of chemical surface preparations that do a very effective job. For long times, approximately 500 hours and greater at temperatures above 450°F (233°C) in air however, these same preparations fail to perform adequately. It is suspected that the oxides formed on the titanium surface undergo changes on long term heat aging that cause failures seemingly in the oxide layer.

Needless to say that when producing adhesive bonded, flight hardware, whether primary or secondary structure, great care is taken to produce the best that the state of the art of technology is capable of producing. This usually requires the development of and strict adherence to detailed and comprehensive materials and process specifications as well as end item non-destructive inspection.

The goal of the surface preparation procedures is to form on the metal adherend an oxide film or other chemical conversion coating that will be stable during the useful life of the adhesive bonded structure.

2.2 Background

The repair of aircraft structural components may be separated into three distinct categories, at least as far as U.S. Air Force operational procedures are concerned. These are the depot, field (base) and battle operational levels. At the depot level at least within the USAF, Navy and the Army extensive and sophisticated capabilities exist, including trained and experienced personnel for the adhesive bonding repair of aircraft structures. All of the facilities available in a modern manufacturing facility, exist at the depot level, and can be brought to bear to solve problems encountered in the repair and rework of aircraft components via adhesive bonding procedures. Thus at the depot level one will find autoclaves, bond form tooling, refrigeration for film adhesives, nondestructive inspection and machine shop as well as engineering laboratory support. At the field or base level (not a forward base under battle field conditions) autoclaves, bonding form tooling and engineering laboratory support would be non-existent, as would be a supply of film adhesives or prepreg fabrics. State of the art adhesive films and prepreg fabrics are difficult to purchase in small lots and have relatively short storage lives (6 months or less) even under refrigerated conditions. There may be many instances, however, where the base level use of film adhesives may be feasible. In these instances, however, reliable storage facilities must exist as well as the means for requalification of the adhesive once it has passed its specification storage life. Machine shop support would exist as well as heat, vacuum pumps and the various bagging materials and supplies that do not have limited storage lives. Trained and resourceful personnel would be available for repair operations, however, bonding experience may be limited. Surface preparations for metals, other than those that could be prepackaged for emergency use, would not be available. It is possible that field level repair carts would be available for use depending upon the particular aircraft system involved.

At forward bases (field level) operating under battle conditions only the most rapid repair procedures will be given consideration. Thus mechanical fastening will be the primary method of repair in this situation possibly backed up by rapid room temperature curing adhesive bonding systems that do not require chemically prepared surfaces to provide high performance bonds.

In our coverage of the surface preparation of metal for bonding we will consider both those surface preparations available at the depot as well as those potentially available at the base.

The technical literature on the pre-bond surface preparation of both aluminum and titanium is extensive and at the same time conflicting. Numerous techniques are available for both classes of metals all with data to substantiate the claims that are made. Considering that state of affairs then this chapter will discuss only those preparations that have found commercial scale utility in the aerospace industry.

2.3 Structural aluminum alloys

Extensive effort has been put into the development of chemical treatments by which we may form a strong stable oxide layer on the surface of the structural alloys utilized in the fabrication of bonded aircraft components. There is more theoretical as well as engineering information available, on adhesive bonding, for aluminum than for any other class of alloy materials [9, 11].

The Air Force under its "PABST" [5, 12] Production Adhesive Bonded Structure Technology, advanced development program thoroughly evaluated a number of surface preparation procedures for aluminum alloys. The testing completed under the PABST program was very comprehensive. The data generated showed conclusively that the "PAA" phosphoric acid anodize surface treatment developed by the late Bert Bethune and his colleagues at the Boeing Corp. is the most durable surface for bonding aluminum alloys presently available. Chromic acid anodize and the optimized FPL etch, however, were close seconds to the PAA treatment [13]. As a result of this evaluation under the PABST program the Air Force has equipped all of its depot centers involved in structural adhesive bonding with the ability to process aluminum with the PAA process. Procedures have been developed to perform the PAA treatment to components while still on the aircraft. This "PANTA" non tank anodize process developed under contract for the Air Force makes it possible to utilize the PAA process in the performance of repair procedures where it may not be desirable or even possible to remove the component from the aircraft and immerse it into the PAA tanks.

The PAA process is preferred at the depot level primarily due to its outstanding performance, simplicity, and ease of disposal of used acid solutions. Extensive amounts of data are now available with various primers and adhesives utilized to bond PAA treated aluminum. The PAA process has been well characterized and is finding extensive use in the aircraft industry. The oxide layer formed by the PAA process is very fragile and porous and must be sealed by an application of corrosion inhibiting primer to prevent the occurrence of inadvertent mechanical damage or contamination and to obtain optimum performance from the system. At the depot level then, repair, rework or remanufacture of a component as the case may be is performed utilizing materials and processes that match as closely as possible those utilized in the original production of the article. The component is nondestructively inspected to determine the extent of the damage or corrosion and the repair/rework procedure tailored accordingly. Once the part has been disassembled and new components substituted for the damaged components and prefitting has been accomplished the PAA process is started. The PAA process is described in the Society of Automotive Engineers (SAE) Aerospace Recommended Practice (ARP) document ARP # 1575. This extremely durable surface treatment is accomplished as follows [14]:

Solvent degrease
Alkaline wash
Deoxidize
PAA

2.4 Phosphoric acid anodizing

The PAA process is characterized by its simple chemistry, room temperature requirements, low electrical needs, environmental acceptability and lack of really critical process parameters. The anodize tank solution nominally contains about 13.5 ounces (0.383 Kg) of concentrated phosphoric acid per gallon (3.785 L) distilled water.

The conditions utilized to form the anodize are nominally 10 volts DC for about 23 minutes at about 80°F (27°C). Upon completion of the PAA, rinsing should take place immediately in order to prevent the resolution of the anodize layer that has been formed. As previously mentioned the anodize layer produced by the PAA process is many times thicker than that produced by the FPL process and is easily damaged due to its columnar, porous structure. Once produced the PAA must be protected with a corrosion-inhibiting primer as quickly as is practicable. The PAA process has been demonstrated to be easily controlled, highly reproducible and to have few if any really critical process variables. For the sake of quality control one most certainly should control bath temperature, bath composition, voltage and time in the bath.

There are always cases in the repair of an article where the use of any chemical surface preparation is prohibited due to the fact that chemicals may migrate into areas where they cannot be readily removed. An example might be migration into honeycomb cells adjacent to the area being repaired. The inability to remove the acids would in all probability cause a site for future failures to be formed.

Structural bonding repair of aluminum components at the base level are both limited and complicated by the factors previously mentioned. These are the lack of film adhesives, chemical surface treatment facilities and experienced personnel.

Structural bonding repair at the base level will be performed using nonstandard surface preparations and two-part prepackage repair kits containing resins, curing agents and perhaps cloth or preprimed aluminum patches for use in the repair process.

Prebond metal surface treatments at the base level will consist of:

Solvent degreasing
Mechanical abrasion
Degreasing primer

The components to be bonded will be solvent degreased by using hand procedures. Mechanical abrasion using scotchbrite type pads will then be accomplished. Next a degreasing primer such as the 3M 3911 degreasing primer is applied. After removal of the 3911 residue again by hand the area to be bonded is primed by a corrosion-inhibiting bonding primer which is then heat cured using hot air or infra red. The prebond treated components are now ready for adhesive application, assembly and cure. The cure will be accomplished utilizing non-autoclave procedures such as vacuum bagging and electrically heated blankets.

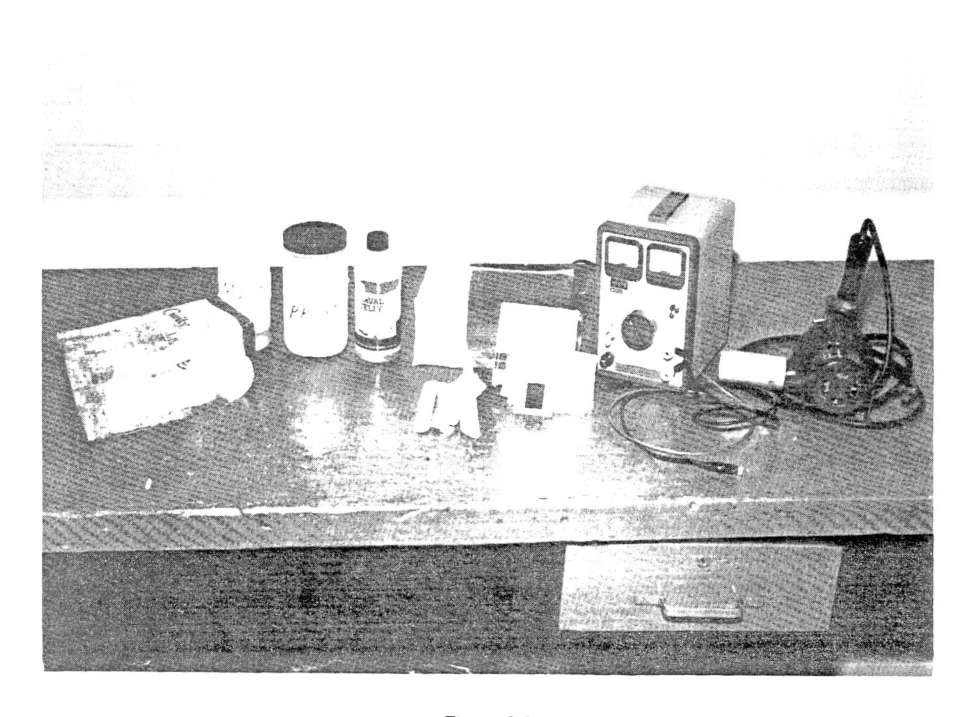

Figure 2.1.

Repairs performed at the field level while not considered to be temporary will most probably be reaccomplished at the depot during the next scheduled overhaul period for the aircraft, where proper surface preparation techniques are available. Bonded repairs performed under emergency conditions will be considered temporary, to be replaced at the next opportunity. Such repair will be performed with mechanical abrasion and a solvent wipe as the surface preparation. Adhesives to be utilized will include the high performance, rapid curing acrylic materials now available.

Laboratory evaluations have shown these adhesives to be very tolerant of metal surface preparation and to develop respectable mechanical properties after a cure of one hour at room temperature.

It is most probable that the emergency repair procedure will combine adhesive bonding and mechanical fastening.

An evaluation of several of the newer rapid curing adhesive bonding systems produced the following preliminary information:

Test temp Adhesive	Room temperature 180°F (82°C)				
	Surface prep I		Surface prep II		
	Shear strength	Peel strength	Shear strength	Peel strength	Shear strength
A	3710 psi (26 M Pa)	29 lbs/in (5.1 KN/M)	3750 psi (26 M Pa)	33 lbs/in (5.8 KN/M)	1420 psi (10 M Pa)
B	2270 psi (16 M Pa)	13 lbs/in (2.3 KN/M)	2120 psi (15 M Pa)	16 lbs/in (2.8 KN/M)	1700 psi (12 M Pa)

cure 4 hrs at room temperature

180°F (82°C) test after 10 minutes 180°F (82°C)

The adhesives were both of the acrylic type and the surface preparations consistent with battlefield condition limitations. Testing of the durability of these adhesives is now underway.

There are three surface preparation techniques that are strong candidates for field or base level metal (aluminum) pre bond surface preparation. All of these treatments, however, pre suppose the availability of a certain amount of materials and equipment in the form of prepackaged kits, and personnel with enough training to be familiar with the processes to be used. Such a pre-supposition may have a relatively low probability of occurrence in many instances that can be conceived [15, 16].

The first of these is the non tank anodize process (PANTA) developed for the U.S. Air Force by the Boeing Corp. In the PANTA process gelling agents are added to the phosphoric solution to allow its use on vertical or the underside of horizontal surfaces. A glass cloth is saturated with the gelled electrolyte solution and is placed on the metallic area to be anodized. A metallic screen (stainless steel or lead) is placed over the cloth and a barrier film of polyethylene, etc placed over the screen. The electrical hook up is the same as if in a tank, with the surface to

Figure 2.2.

be bonded, the anode. Care must be taken to work all the air out of the contact area between the metal surface and the saturated glass cloth carrier. The surface, after anodizing must be thoroughly rinsed and dried prior to the priming and bonding processes. This is a preferred process due to the fact that the PAA solution is relatively non-toxic and non-corrosive. Nevertheless protective eyeware or face marks, gloves and aprons should be utilized to protect personnel from contact with the electrolyte. In this field level process precleaning with fine grit paper or scotchbrite material and solvent wiping may be substituted for a chemical deoxidizing treatment.

The following is a listing of needed materials and equipment for the successful field application of the PANTA procedure:

Required Materials/Equipment:

1. M.E.K. trichloroethane or equivalent
2. Nylon abrasive pads or equivalent
3. Clean gauze
4. Phosphoric acid – 12% solution
5. "Cab-o-Sil" thixotropic agent
6. Stainless steel screen – similar to window screen mesh
7. Rectifier capable of supplying 6 volts D.C. with current density of up to 7 amps/ft (63) amps/m^2

Figure 2.3.

8. Distilled or deionized water
9. Polarizing filter (for inspecting phosphoric acid anodize – when Surface Inspection Tool is available, this is not necessary)
10. Plastic spatula for kneading cheesecloth/gauze – acid get stack, to remove air from metal surface.
11. Air to force dry at 140 to 160°F (60 to 71°C)

Figure 2.1 shows several of the above items that were utilized in an actual field demonstration. Figure 2.2 shows the phosphoric acid solution in "gel" form being applied to the lower surface of an aircraft control component. Figure 2.3 shows the application of the stainless steel screen and electrical hookup [17, 18].

A much older process that has been utilized with some success is the "Barium Sulfate" thickened FPL etch treatment. In this process the FPL etch is thickened with finely divided barium sulfate until it reaches the proper consistency to be handled and not to run off or spread on vertical or horizontal surfaces. This material is very corrosive due to the free sulfuric acid present and is toxic due to the dichromates present. In addition the treatment must be accomplished hot (150°F) (66°C) to obtain acceptable durability in the subsequently bonded components. This is not a preferred method due to the difficulty of removing the sulfuric acid and the toxicity of the chromates present, and the danger of the migration of the sulfuric acid into areas where it may be difficult or impossible to remove [19–21].

A newer process and one without systematically obtained data to substantiate its effectiveness and durability is described by Dr. Leo Windecker. This process involves the mechanical abrasion of the adherends to be bonded using a fine grit paper or scotchbrite pad under a protective layer of the adhesive (liquid in this case) resin in combination with a silane wetting agent. The theory behind this process involves the formation of relatively long lived free radicals on the metal surface by the mechanical abrasion under the protective blanket of the resin material. The though here is that primary bonds can be formed and durable structures will result. While theoretically sound little data are available to describe its performance. It is an attractive process that will require an experimental program to reduce the theory to practice and to optimize the many process variables that are involved.

2.5 Chromic acid anodizing

While chromic acid anodizing is still a commercially utilized process no time will be spent discussing its pros and cons here due to the fact that the "PABST" work showed the PAA process to be simpler and far superior in performance and also that the industrial disposal of spent chromate solutions is a difficult and expensive job. It is the author's opinion that "where feasible" the industrial use of environmentally acceptable chemical processes is to be preferred [22, 23].

2.6 Titanium alloys

The prebond chemical treatment of titanium alloys for acceptable bond durability has in the past required very energetic combinations of acids, or bases [24]. Such combinations include, nitric-hydrofluoric, phosphate modified hydrofluoric, sulfuric acid-peroxide, alkaline-peroxide or hot alkaline etches. While these treatments may or may not be acceptable for factory or depot usage they are certainly out of the question where base or field level surface treatment of titanium alloys is necessary. The Boeing Corp. has developed a modified PAA treatment for titanium alloys that may be very acceptable for field or base use provided the proper materials and training can be made available to the personnel responsible for performing the procedures. Fortunately base or field level structural bonding of titanium alloys will not have to be accomplished due to the fact that field level surface preparation is not feasible and also that so little Ti is utilized in typical applications, so the problems here are for the most part academic [25–28].

2.7 Summary

The prebond surface preparation of metallic adherends is the most critical step in the adhesive bonding process. If not done correctly all else is for naught. Studies have shown that variations in curing conditions and wide variations in adhesive chemistry are not as critical as is the pre-bond surface preparation given to the metal.

Combinations of heat, moisture and stress as obtained in the cantilever type loading obtained in the wedge opening test quickly reveal any weaknesses in the metal surface preparation procedures.

The relatively recent development of very sensitive analytical tools for studying the physical and chemical characteristics of surfaces and their application to adhesive bonding processes has rapidly pointed the way to optimization of surface characteristics.

Structural adhesive bonding processes are characterized by the high level of technology required when compared to those utilized in mechanical fastening processes. Since many components cannot be repaired by mechanical fastening and adhesive bonding offers many structural advantages, we are concerned with how to do the best job of adhesive bonding that we are able with the materials, processes and equipment (technology) available to us.

For the bonding of aluminum alloys the PAA processes are favored due to the high durability obtained, simplicity of processing and environmental acceptability.

The prebond surface preparation of titanium alloys requires highly energetic chemicals suitable for use at the factory and depot levels but not for use at the base or field levels where the needed equipment and trained personnel may not be available. The proprietary Boeing PAA for titanium alloys may be an answer to this situation.

Finally much new and existing work is now on going to develop and introduce new technology into the area of depot, base and field bonded repairs. This work is focused on the development of new materials, processes and equipment with emphasis on new and novel technology for the purpose of repairing both bonded metallic and composite structural components.

References

[1] Cagle, C. V., Adhesive Bonding Techniques and Applications, McGraw Hill (1968).

[2] Titanium Surface Preparations, AFML-77-146.

[3] Houwink, R. and Salemon, G., Adhesion and Adhesives, Volume 2, Elsevier Publishing (1967).

[4] Reinhart, T. J., Evolution of Structural Adhesives, in 'Structural Adhesives and Bonding Technology Conferences, Conference on Structural Adhesives and Bonding, (1979).

[5] Primary Adhesively Bonded Structure Technology, AFFDL-TR-77-107.

[6] Leaves and Natatojan, Systems Approach to Bonding.

[7] Bikales and Norbert, Adhesion & Bonding.

[8] Adhesives in Modern Manufacturing, Society of Manufacturing Engineers.

[9] Smith, T., Surface Treatment for Aluminum Bonding, Rockwell Science Center, Report SC5180-17FTR (1979).

[10] Bikales, N. M. (Editor), Adhesion and Bonding, New York, Wiley International (1971).

[11] Smith, T., Surface Science 55, 601, (1976).

[12] AFML-TR-74-73, Part II, Oct. (1975).

[13] Hoisch, H.L., Report No. MT-77-021, May (1978).

[14] DeLollis, N., Adhesives For Metals: Theory and Technology, New York, Industrial Press (1970).

[15] Hochberg, M.S., Methods of Bonding Cryogenic Structures, Report No. F-743, 1 Sept (1967).

[16] O'Brien, R.W., Strength Tests of Titanium Honeycomb Panels, Report No. 604-175, 30 Jan. (1968).
[17] U.S. Patent No. 3,804,730, Control of Electrolytic Coloring of Chromium-Containing Alloys, T.E. Evans, W.H. Sutton, Intern'l Nickel Col, Inc. New York, NY.
[18] Moji, Y., Preparing an Environmentally Stable Stainless Surface for Bonding, U.S. Patent No. 4, 064, 020, The Boeing Company, Seattle, WA.
[19] Hudson, W.R., *J. Vac. Sci. Tech.*, 14, 286, (1977).
[20] Beck, W., Bockris, J., McBreen, J. and Nanis, L., *Proc. Royal Soc. London*, A290, 220, (1966).
[21] Smith, T., *J. Applied Phys.*, 46(4), 1553, (1975).
[22] Pochily, T.M., Process for Anodizing Titanium, Technical Report WVT-6605, Benet Laboratories, Watervliet Arsenal, Watervliet, New York, April (1966).
[23] Walter, R.E., Titanium Processing Technology – Adhesive Bonding, Report No. A0180-2, 3 Mar. (1970).
[24] Voss, D.L., Adhesive Bonding of Titanium, Report No. A0449, 1 Aug. (1970).
[25] Monroe, R.E., *et al.*, Joining of Titanium, TR DMIC240, Defense Metals Information Center, Battelle Memorial Institute, Columbus, OH.
[26] Investigation of Metalbond 329 Strength Reduction in Salt Spray Environment when Bonding Titanium, MCAIR Technical Memorandum 256.756.
[27] Nondestructive Test Methods, MCAIR Engineering Study Authorization Interim Report No. AO121, 30 Jan. (1970).
[28] Kieth, R.E., *et al.*, Adhesive Bonding of Titanium and its Alloys, NASA Tech. Memorandum NASA TM-X-53313, 4 Aug. (1965).

L.J. HART-SMITH

3

Design and analysis of bonded repairs for metal aircraft structures

3.1 Introduction

Research conducted under the Primary Adhesively Bonded Structure Technology (PABST) program proved that adhesive bonded structures are far more tolerant of quite large flaws than had previously been believed. The PABST fuselage was made on development tooling rather than on production tooling which could have been refined after manufacturing began. Consequently, every panel was accepted for the structure and most of the flaws were left unrepaired, to be monitored during the testing. Because the PABST fuselage was not a flawless structure, the success of the test program was of even greater significance than it would have been for some perfect laboratory test coupon. The bond flaws simply did not grow, which gives great confidence in the reliability of adhesively bonded structures. Even the slow-cycle testing under extreme environments and loads did not damage the bonds in test coupons.

It was therefore concluded that many bonded repairs made in the past were simply unnecessary. Worse, in many cases, the repair served only to reduce the service life by providing additional paths for moisture to ingress. Had such flaws been unrepaired, the structures would have been just as strong and lasted much longer.

With the introduction of male bonding tools halfway through the program to supplement the original female bonding tool, the manufacturing cost of each panel was reduced dramatically. In addition, the number and size of the bond flaws was also decreased. Consequently, it does not follow that costs can be reduced by increasing the acceptable flaw sizes. The subject of structurally acceptable bond flaw sizes simply cannot be considered in isolation from the economic factors which tend to call for still smaller flaws. This does not mean that most damaged structures should be left unrepaired. Rather, the repair criteria have been inadequate and, insufficient use has been made of the analyses of structural bonded joints in the original structures. After all, a bonded joint used in a repair should be as similar as possible to the bonded joint in the original structure.

3.2 Design of adhesive-bonded repairs in thin sheet metal contruction

Experience with adhesive-bonded joints gained in the PABST program can serve as the basis for the design and fabrication procedures for many bonded repairs. The pertinent information is presented here and the reader can find details of the derivations in other sources [1–4].

Figure 3.1 shows the overlaps used for the adhesively bonded splices in the aluminium structure of the PABST fuselage. These can be approximated by the simple rule that the overlap should be about 30 times the central skin thickness for double-lap joints. For unsupported single-lap joints, the corresponding overlap is 80 times the thickness, to protect the adherends against the bending moments set up in the adherents and to protect the adhesive against the peel stresses which peak at the ends of the overlap. These same proportions would be equally useful for bonded repairs. The methods used to establish these overlaps can be extended to other types of adhesive and adherend materials, as for fibrous composite structures [5].

The tapering of the splice plate referred to in Note 2 of Figure 3.1 is shown in Figure 3.2. Tapering is needed for the thicker adherends to prevent a premature failure from induced adhesive peel stresses before the adhesive had developed its full shear strength. The same tapering is also appropriate for single-lap joints. The precise amount of tapering is not critical [6]. The reason for this is that the "other" end of the joint, in the middle of the splice shown in Figure 3.1, then governs the joint strength. The local thickening of the adhesive layer must be restricted so that the adhesive does not run out under capillary action and the adherend must not be excessively thinned lest the edge of the splice strip become prone to damage in handling. But otherwise, it is almost impossible to do any harm by applying the

CENTRAL SHEET THICKNESS t_i (IN.)	0.040	0.050	0.063	0.071	0.080	0.090	0.100	0.125
SPLICE SHEET THICKNESS t_o (IN.)	0.025	0.032	0.040	0.040	0.050	0.050	0.063	0.071
RECOMMENDED OVERLAP[1] ℓ (IN.)	1.21	1.42	1.68	1.84	2.01	2.20	2.39	2.84
STRENGTH OF 2024-T3 ALUMINUM (LB/IN.)	2600	3250	4095	4615	5200	5850	6500	8125
POTENTIAL ULTIMATE BOND STRENGTH (LB/IN.)[2,3]	7699	8562	9628	10,504	10,888	11,865	12,151	13,910

[1] BASED ON 160°F DRY OR 140°F/100-PERCENT RH PROPERTIES NEEDING LONGEST OVERLAP.

VALUES APPLY FOR TENSILE OR COMPRESSIVE IN-PLANE LOADING. FOR IN-PLANE SHEAR LOADING, SLIGHTLY DIFFERENT LENGTHS APPLY.

[2] BASED ON -50°F PROPERTIES GIVING LOWEST JOINT STRENGTH AND ASSUMING TAPER OF OUTER SPLICE STRAPS THICKER THAN 0.050 IN. STRENGTH VALUES CORRECTED FOR ADHEREND STIFFNESS IMBALANCE.

[3] FOR NOMINAL ADHESIVE THICKNESS η = 0.005 IN. FOR OTHER THICKNESSES, MODIFY STRENGTHS IN RATIO $\sqrt{\eta/0.005}$.

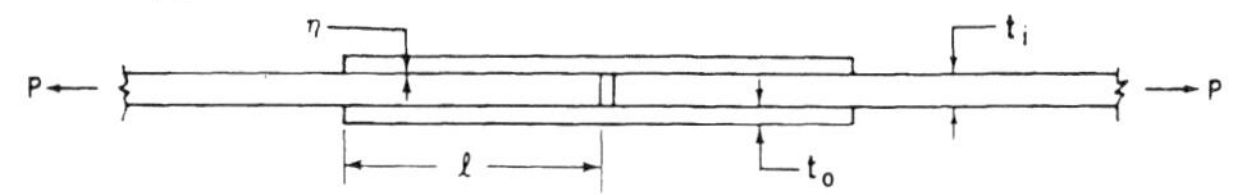

Figure 3.1. Design overlaps used for PABST skin splices.

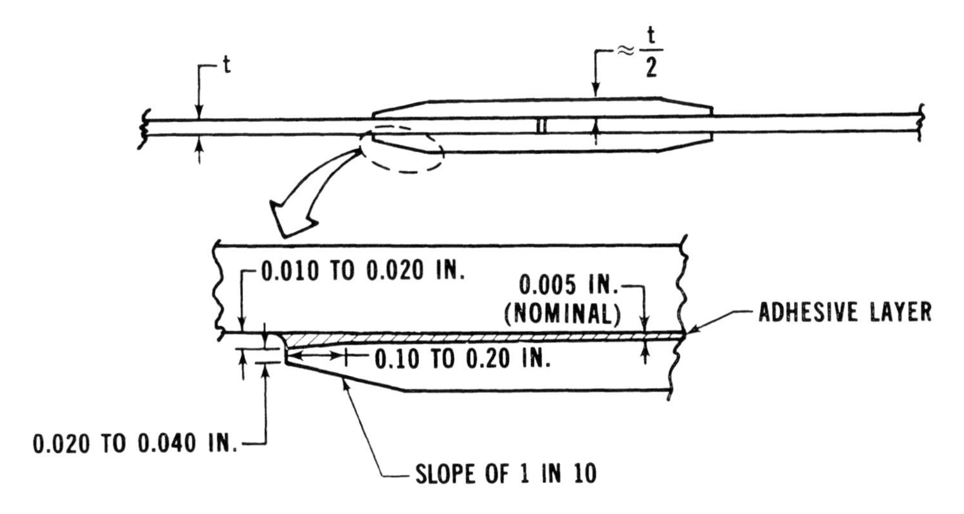

Figure 3.2. Tapering of edges of splice plates to relieve adhesive peel stresses.

techniques shown in Figure 3.2. Of course, they are unnecessary for the thinnest of the adherends, as indicated in Figure 3.1.

The proportions shown in Figure 3.1 can be applied both to stiffened sheet construction and to the facings and edge members of honeycomb panels. When a flush surface is not required on one side of a honeycomb panel, it can be repaired by the simpler technique shown in Figure 3.3. The overlap can then be 50 to 60 times the sheet thickness since the eccentricity in load path is then reacted by the honeycomb instead of causing bending deflections of the skins. The single-lap splices have only one bond surface to transfer load through the adhesive, so the upper limit on thickness was set at 0.071 inch during the PABST program (instead of the 0.125 inch in Figure 3.1).

The overlap and details shown in Figures 3.1 and 3.2 are close to optimum for most ductile structural adhesives and are conservative for the brittle high-service-temperature adhesives. They have also been approved by skeptics who size bonded joints on the basis of whether fasteners can be installed if the adhesive should fail. They could also suffice for room-temperature-cured adhesives, even though they can only be used up to a lower maximum service temperature. Many room-temperature-cured adhesives become soft at a lower temperature than can be withstood by heat-cured adhesives. This only mild sensitivity of optimum joint proportions to the type of structural adhesive used may seem incongruous in relation to the wide disparity between the so-called adhesive shear strengths measured on the standard lap-shear test coupon. The reason is that the optimum overlap is usually set by the hot, wet service condition and most adhesives have similar stress-strain curves at their maximum service temperatures, even though those temperatures vary with the adhesive type and the glass transition temperature.

If space limitations for repair dictate lesser overlaps than shown in Figure 3.1, other methods [2] can be used to justify smaller overlaps for the more brittle adhesives. However, the brittle adhesives, such as FM-400, M-329, and HT-424, have much lower strain energies to failure than do ductile adhesives like FM-73, Redux 775, EA-9628, FM-123, AF-126, FM-1000, EA-951, and others used extensively on subsonic transport aircraft. Consequently, the maximum thickness of members that can be adhesively bonded with the simple joints shown in Figure 3.1 will be reduced for brittle adhesives.

3.3 Residual strength of flawed or damaged adhesive bonded joints

Adhesive-bonded repairs are, of course, needed in the case of gross structural damage. However, with a local disbond, testing on the PABST program clearly indicated that adhesive bonds are far more tolerant of flaws and porosity than had been believed hitherto. The fourth and fifth lines in Figure 3.1 reveal that, for the thinnest aluminium alloy sheets, the potential shear strength of the bond exceeds the strength of the aluminium sheet outside the joint by a factor of about 3 to 1. For the thickest sheets shown, however, that factor is less than 2 to 1. It is evident that, for a still greater thickness (actually, about 0.125 inch), the shear strength of the adhesive-bonded joint will be less than the load which the members outside the joint could transmit. This indicates that there is a certain skin thickness for each adhesive at which the bond shear strength precisely balances the adherend strength. Under such circumstances, the slightest imperfection in the adhesive bond could result in a situation in which the bond would unzip throughout its entire area because the adherends could transmit just enough load to strain the adhesive beyond its ultimate strain capacity. This is explained in Figures 3.2 to 3.4 of another work [7]. Obviously, in bonding still thicker aluminium adherends, mechanical fasteners would be needed to achieve any fail-safety at all. The effectiveness of fasteners in such a bonded/bolted configuration is discussed elsewhere [8].

The thinnest aluminium adherends in the table in Figure 3.1 are simply incapable of inducing enough load to strain the adhesive to failure, except for two conditions. First, if the adherends are left under a sustained load sufficient to yield them, a progressive bond failure will follow, as shown in Figure 3.4. When tested under rapidly applied loads, as induced by gusts rather than cabin pressure, the same joints failed consistently outside the joint at much higher loads without damage to the adhesive. The necking down shown in Figure 3.4 continues until the remaining bond area is just sufficient to resist the applied load. By analogy, the joint could withstand quite a large portion of defective or damaged bond and still be stronger than the adherends. In other words, the excess strength of the bond over the adherends in Figure 3.1 represents a measure of damage tolerance for the adhesive-bonded joints. This cushion must never be allowed to decrease to zero. It should never be allowed to be reduced below a 50 per cent excess strength, no matter how thick the members being joined are. That is why the cutoff in Figure 3.1 is set at 0.125 inch instead of 0.188 inch, the next gauge up. The same considerations should also apply to the repair of damage which occurs in aircraft service or of defects which occur during manufacture. The bonded repairs must be

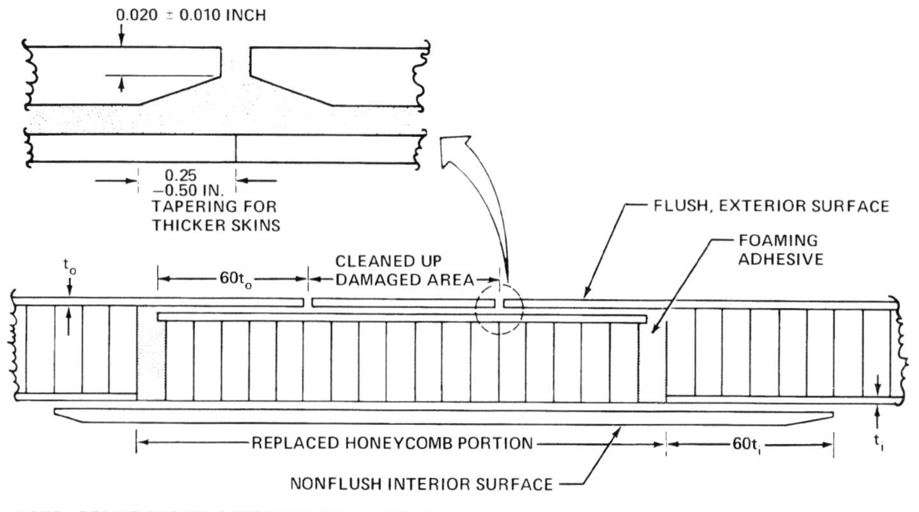

Figure 3.3. Repair of skin damage to bonded honeycomb panels.

at least as strong as the adherends, regardless of the nominal loads on the intact or repaired structure, because any lesser repair could act as a weak-link fuse in the event of subsequent damage to an adjacent area.

The damage tolerance of properly proportioned adhesive-bonded joints is indicated in Figures 3.5 to 3.8, which show the adhesive stress distributions in an intact PABST fuselage splice and three joints with different bond flaws. None of the flaws shown decreases the joint strength or increases the maximum adhesive shear stress and strain, even though they would all be "repaired" according to prevailing acceptance criteria. Actually, only the defect in Figure 3.6 needs any repair, by a good rubbery sealant rather than by an adhesive, to prevent the initial damage from spreading due to the freeze/thaw cycles which would follow if moisture were allowed to penetrate through the exposed crack at the edge of the splice plate.

The absolute maximum size of local bond flaw which can be tolerated is easy to establish in terms of the joint strength versus overlap characteristic in Figure 3.9. If the effective overlap were to be reduced from its original design value to the transitional value at the bend in the curve, the adhesive would become the weak link instead of being protected by the limiting strength of the adherends outside the joint. A minimum effective overlap would be 50 per cent greater than that, corresponding with a 50 per cent strength margin. It is immaterial how the remaining effective bond area is distributed across the overlap. The strength would be essentially the same. Thus, bonded joints with quite large flaws can be left unrepaired. However, even lesser flaws which are exposed to moisture should be sealed.

There is a very strong consideration based on service experience which should discourage all unnecessary repairs of flaws or damage to adhesive bonded joints.

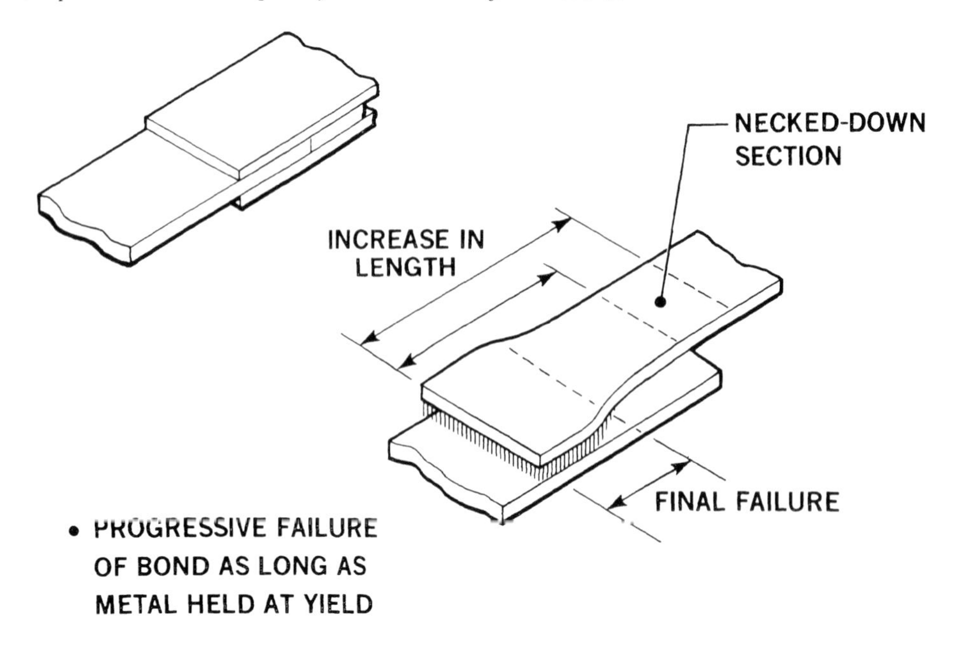

Figure 3.4. Effect of yielding of metal.

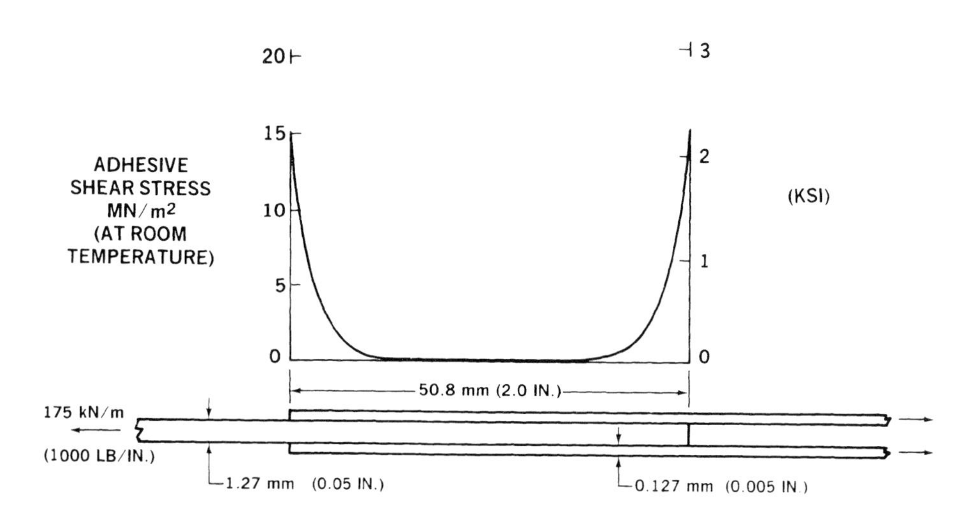

Figure 3.5. Adhesive shear stresses in bonded joints.

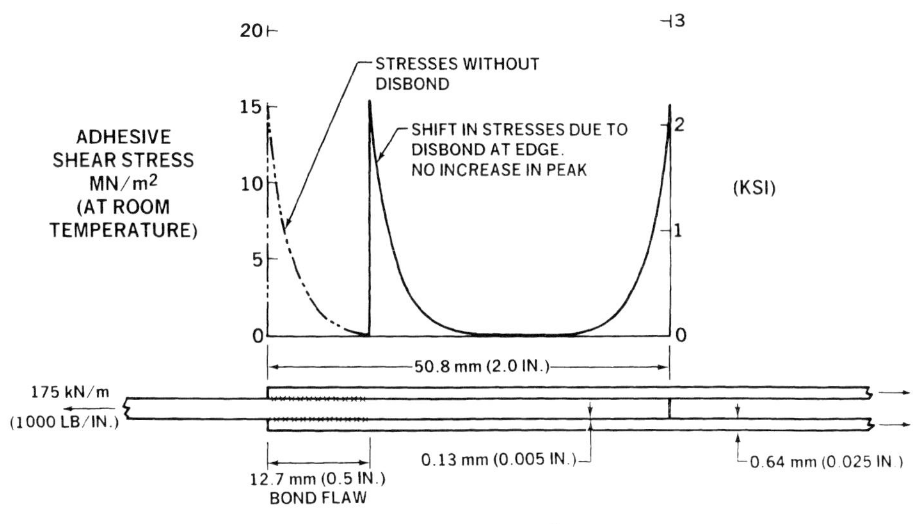

Figure 3.6. Adhesive stresses in flawed bonded joints.

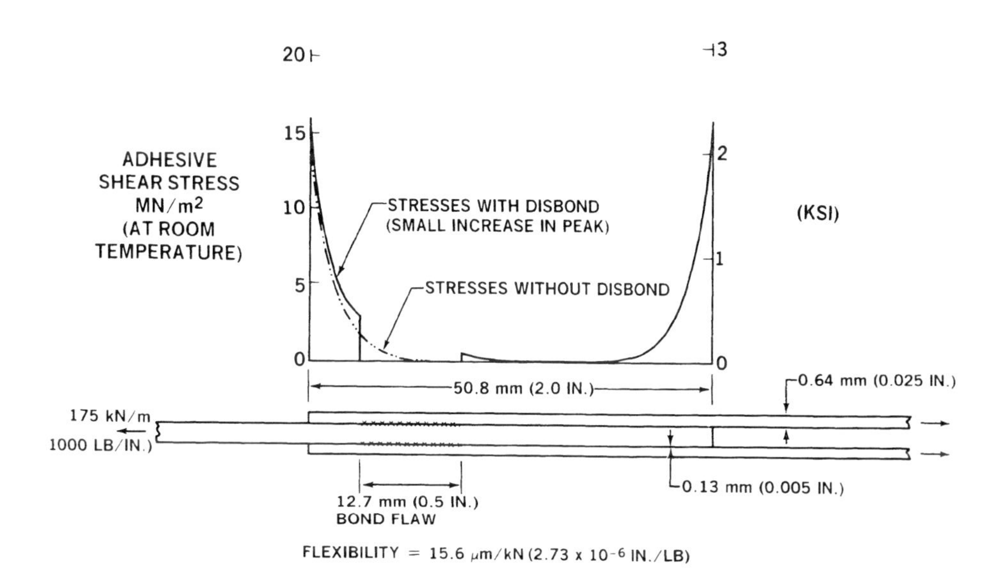

Figure 3.7. Adhesive stresses in flawed bonded joints.

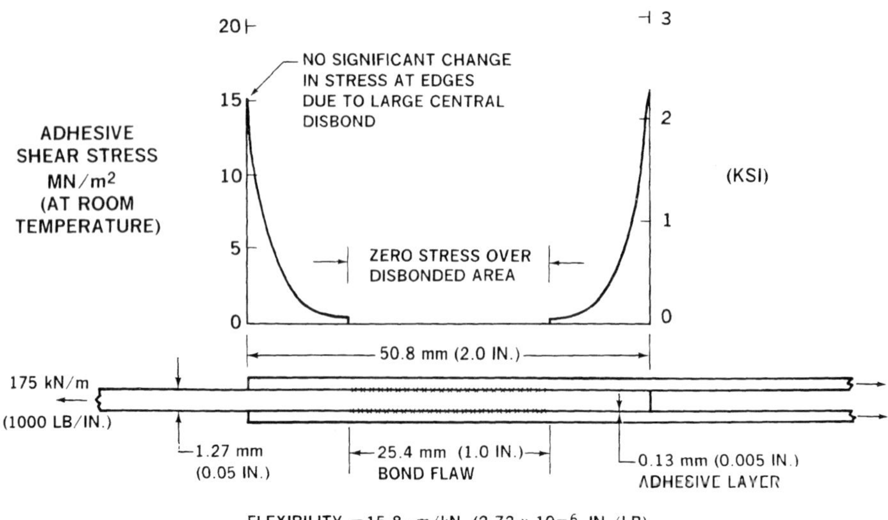

Figure 3.8. Adhesive stresses in flawed bonded joints.

Unless the flaw is exposed to an edge of the bonded overlap, it can be repaired only by drilling holes to inject the repair adhesive. Those holes must break the surface anodize, or etch, and any primer, which, in turn, destroys the primary resistance of the bonded joint to hostile service environments. It is ironic that many in-service repairs to bonded structure would never have been necessary if the initial flaws had only been left alone at the factory prior to delivery. The corrosion on the bonded panel shown in Figure 3.10 which followed the disbonding due to moisture attacking the adhesive-to-aluminium interface is typically initiated at the edge of a panel

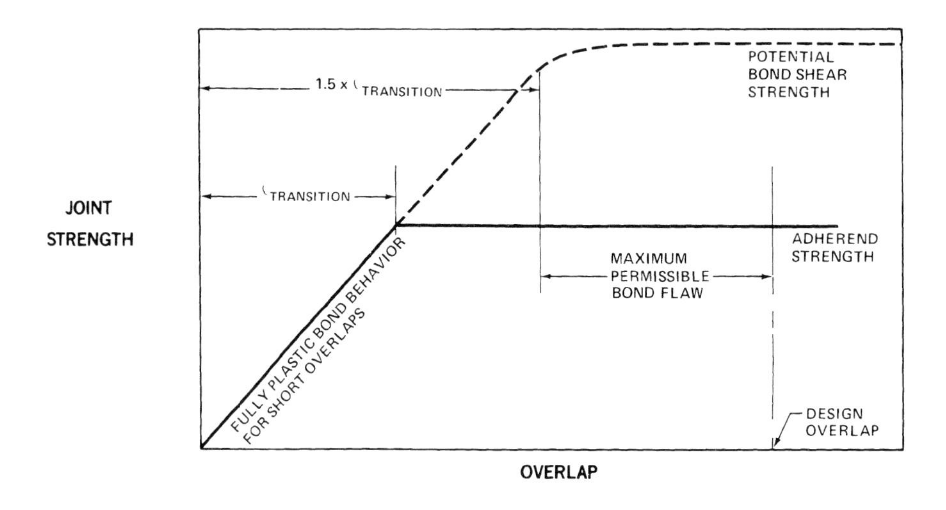

Figure 3.9. Establishment of maximum permissible adhesive bond local flaw sizes.

Figure 3.10. Corroded bonded joint.

trimmed to fit on assembly or arises from a fastener hole drilled after bonding. In both cases, the surface protection, meager as it may have been, was completely destroyed before such components ever entered service.

3.4 Acceptance criteria for bond flaws and damage

The preceding section suggests that the existing acceptance criteria for bond flaws could be liberalized, on the basis of structural strength considerations. However, minimization of both the manufacturing and maintenance costs of adhesive bonded panels tends to result in fewer and much smaller bond flaws than are accepted today.

Suppose, for example, that a bond flaw is detected by inspection during the 20th year of the life of an aircraft. Obviously, if it really had just appeared and was actively growing, it would need a prompt repair, with the aircraft out of service. To justify that the flaw does not need repairing, 19 years of prior records would have to be produced in order to show that the flaw has been there all along and has not grown in size. Without such mountains of documentation on each bonded panel on every aircraft, it would have to be assumed that such a flaw was new and warranted repair. So it is certainly not in the interest of the airlines or the military to encourage a situation which allows bond flaws to exist, even though statistically it is known that most of them would be structurally harmless.

Even while the bonded panels are being manufactured, substantial savings can be realized by a cost-control policy which discourages the creation or acceptance of flaws. It might seem that costs could be minimized by omitting the close coordination between design and tooling concepts needed to achieve a high-quality bonded structure – and that further savings could be achieved by increasing the tolerances on detail parts and by accepting sloppy workmanship. While that is simply not true, many cost-control policies actually increase the overall costs significantly by minimizing each individual operation as it occurs (frequently only once) without consideration of the effect on subsequent operations (which are usually repeated hundreds or thousands of times). The task of making a structural design and a bonding tool compatible with each other to produce flaw-free assemblies need only be done once. However, the penalty for not doing so is an increased number of flaws to be identified and mapped in every single bonded panel. That causes higher inspection costs, higher rejection rates, and higher engineering salvage efforts. It is too late to build quality into an assembly after it has been bonded together. The money spent on postbonding inspections of significantly imperfect panels would better be spent on improving the design, tooling, or prefit operations to eliminate the need for anything but cursory postbonding inspections on a few randomly selected panels. During the PABST program, it was said that, if the parts fitted together properly before bonding, there was no need to inspect them after bonding. Conversely, if they did not fit together properly before bonding, there was no point in bothering to inspect them after bonding.

The greatest motivation for determining the structural limits on acceptable bond flaws is that, with such knowledge, it would become possible to defer repairs and leave aircraft in service safely until a more opportune time arises to perform such repairs. The economic incentives to eliminate bond flaws before they are manufactured will probably transcend any desire to investigate acceptable bond flaws in aircraft before delivery. Such a philosophy would be consistent with the lack of in-service observation of growth of disbonds under mechanical loads (as distinct from moisture attack, independent of the load intensity).

It should be noted that one source of small bond flaws is inevitable. It occurs frequently in honeycomb structures, but the associated flaws are harmless. It is simply not possible to fit honeycomb cores perfectly in the various corners of the edge closing members. Most of the flaws in such panels occur in these locations. There is usually a very good fit on the adjacent flat surfaces where the load is transferred. A similar situation arises in the bonding of nested angles. The correct solution to this corner condition is to brand a narrow strip, typically only 0.125 inch wide, as not requiring inspection. An added benefit of such an approach is the avoidance of spurious readings of the ultrasonic NDI methods which occur at those corners. Thus, the inspection costs would be reduced much more than the small area no longer being inspected would indicate.

3.5 The pitfalls of life prediction for adhesive-bonded joints

For about a decade, bonded panels with an inadequate surface preparation and moisture-absorbent adhesives have demonstrated an inadequate life. This has been particularly true of panels using clad 7075 aluminium alloys. While many such panels remain in service and some will continue to fail,* the problem has been solved – both by tests on the PABST Program and from the steadily improved experiences of the airlines having bonded panels made with the phosphoric acid anodize which The Boeing Company introduced several years ago. Unfortunately, a legacy of this problem which triggered the concern is known to be incapable of analysis because it is independent of the load level. This research also overlooks the evidence of some 40 years of satisfactory service for bonded panels which are not prone to attack from moisture; it overlooks the tests on coupons cut from the old bonded structure showing no degradation of the adhesive strength, and it overlooks the absence of any progressive fatigue failure of bonded joints in service under mechanical loads.

Attempts to predict the life of bonded joints will result in inferior and heavier bonded structures than those in service today by encouraging the thought that a finite-life bond is permissible.

That would nullify the benefits of the virtually infinite bond life due to relying on the adherends to protect the adhesive and using joint geometries which never permit the weak link to occur in the adhesive. Of course, for such thin structure as control surface skins, the adhesive bond will remain stronger than the skins no matter what analysis method is used to justify the design. In such cases, the harm from the more elaborate bond crack-growth analyses would be confined to the diverted engineering effort which should have been spent on some real area of concern elsewhere in the structure.

The harm done from da/dN calculations for adhesive bonds equivalent to those being introduced for cracks in metal structure would be worse for more heavily loaded structure. Improperly sized bonded joints could even result in catastrophic failures. Since the new damage-tolerance and residual-strength calculations for the growth of cracks in metal structure obviously have considerable merit – even beyond their effect in discouraging the use of the most brittle metal alloys – it would seem appropriate to explain why such methods have no application to well-designed adhesive-bonded joints. The key to the difference is whether redundant load paths are absent or present. The growth of a skin crack in a panel remote from any stiffeners starts off steadily and is quite stable until some critical length is reached. The subsequent fast fracture should be arrested by the stiffeners. The crack growth occurs because there is no alternate load path, so the load is diverted around the crack tip. The crack growth is then controlled and predictable, and can be checked in periodic inspections. However, if the skin does not fast fracture and

*The local repair of disbond damage in panels suffering from the condition shown in Figure 3.10 is futile. When returned to service, the adjacent areas will fail promptly due to further attack from moisture. The only effective treatment is to rebuild the panel using the latest alloys, surface preparations, primers, and adhesives used in current production rather than to the original specifications.

the crack grows further toward and over the stiffeners, the rate of crack growth is reduced and possibly even completely stopped, as with a skin crack over a bonded stiffener. The reason for the crack arrest by the stiffeners is that now the load redistribution due to the skin crack has an alternate path other than around the crack tip. The redundant structure, in the form of the stiffeners, holds the crack tip shut during the stage of the crack growth.

On the other hand, the adhesive in a well-designed bonded joint is always in the benign environment where the adherends restrict the strain on the adhesive, and thereby prevent the growth of any disbonds or any microcracks within the adhesive layer. That is why all tests on bonded-joint disbond growth have shown that a peel load component is necessary if there is to be any disbond growth – there is no alternate load path for peel loads as there is for shear loads.* The ability of bonded joints to redistribute loads across disbonds is well illustrated in Figures 3.5 to 3.8. That situation is mathematically a one-dimensional problem, but much the same relief exists for two-dimensional disbond problems in which a disbond is so large as to divert load around it rather than across it. This two-dimensional disbond problem is illustrated in Figure 3.11. Figure 3.11(a) shows how, in a well-designed bonded joint, any gross disbond will eventually cause a fatigue crack in the skin outside the joint rather than any growth of the disbond inside the joint. In the situation shown in (b), there will initially be some disbond growth because the mechanical fasteners are far more flexible than the adhesive bonds, until the load is transferred by the fasteners across the disbond instead of around it. At that point, any disbond growth stops and the life is set by the growth of any skin cracks initiating from the most highly loaded fastener. While that fastener load obviously depends on the adhesive properties and the size of the disbond, it is obviously independent of the rate of growth of any disbonds. The growth of any initial disbond to the stable size shown would be virtually instantaneous. Figure 3.11(a) also identifies an area needing further research. While the method shown in Figure 3.9 will suffice to determine how large a disbond is permissible and still remain a one-dimensional problem, the two-dimensional situation in Figure 3.11(a) is akin to a skin crack and there should be some evaluations made to relate the initiation and growth of the skin cracks to the size of the gross disbonds. They, in turn, could be used to establish acceptance criteria for tolerable disbonds which would not require immediate repair. The disbond size would be involved in the load redistribution which would induce the skin cracks, but the inspection intervals and repair criteria would otherwise be based entirely on metal behaviour and not on the behaviour of the adhesive bond.

*For the same reason, adhesive-bonded joints are almost always designed to have the primary load transferred by shear. Induced peel loads are minimized, as in Figure 3.2. Even those joints with a pure stress component, as between the skin and frame tees on the PABST fuselage, are proportioned so that the deflections of the metal components restrict the peak peel stresses developed. There are no structural joints in which the primary load is in peel and for which there is no restraint on any crack opening displacement. Even the double-cantilever-beam test coupons used to measure adhesive crack growth under peel loads are displacement-controlled, so that the delaminating force decreases as the crack grows during each cycle. Otherwise, if the test were load-controlled and there were no way to reduce the load quickly enough to arrest the disbond each time it started, all the coupons would fail in a single cycle.

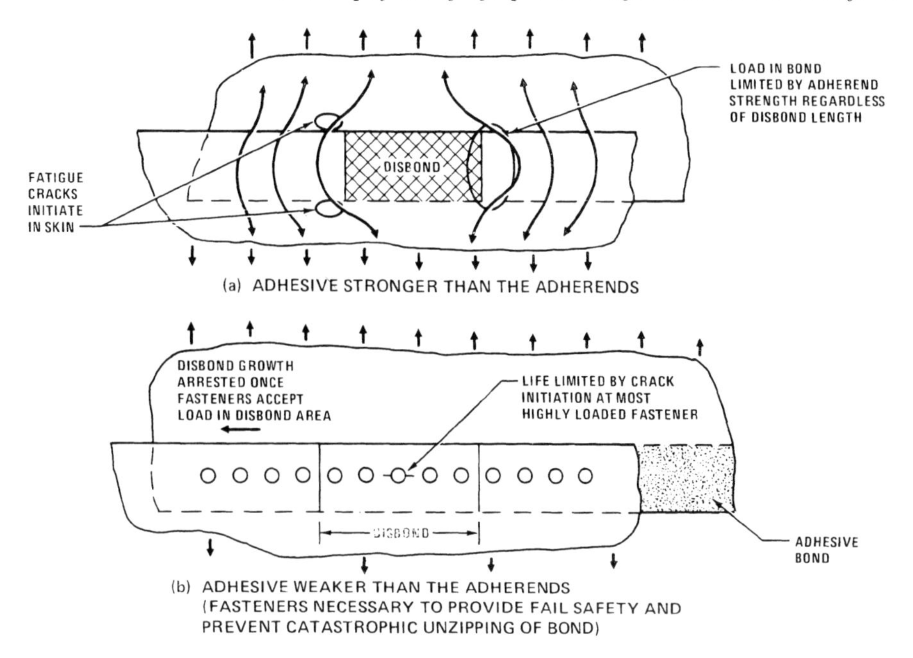

Figure 3.11. Load redistribution around two-dimensional adhesive-bond flaws.

Figure 3.12 summarizes the available experimental evidence on disbond growth in adhesive-bonded structures and highlights the differences evidence about cracks in metal structural components. Apart from indicating the gross difference between behaviour of adhesives under shear and peel loading, Figure 3.12(b) also introduces the one likely source of adhesive disbonds in service – prior fatigue cracks or damage in the metal structure. This subject is explained elsewhere by the author [9]. Briefly, disbonds are far more likely to occur in structure other than

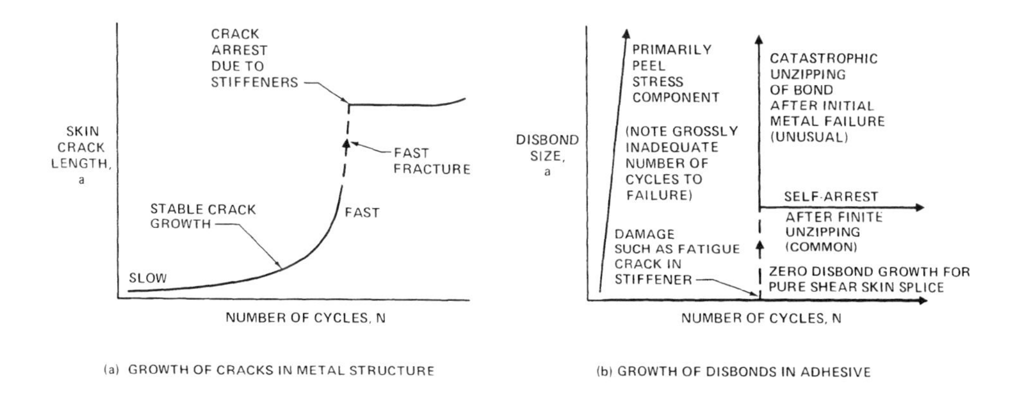

Figure 3.12. Differences between growth of cracks in metal components and adhesive bonds.

bonded splices because the effective member thickness per unit width of bond is much greater for stiffeners, which might be broken abruptly, than for splice plates, which will be tapered smoothly. The consequence of a fatigue crack through a stiffener – emanating from a fastener hole, for example – is most likely to be a fatigue crack in the skin directly in line with the break in the stiffener. The next most likely consequence is stable, self-arresting disbond, provided that the stiffeners have not been designed to be so large and so far apart that the structure should have been riveted instead of bonded. Finally, there are some geometries, such as a multibay skin crack which could occur only in test panels, in which the stiffeners completely disbond before failure. In the disbond behaviours shown in Figure 3.12(b), disbond growth either does not happen at all or it happens so rapidly that there is usually no real interest in establishing the short interval in which the growth occured. Nevertheless, there is one legitimate application of bonded repairs which are known to be so weak that the repair will fail in service. That application – adhesive-bonded boron-epoxy patches to retard the growth of fatigue cracks in thick metal structure – is discussed elsewhere in this volume. The measure of success is prolonging the life of expensive or difficult-to-replace components. Such patches on thin sections or sheets have completely arrested the growth of the cracks in the metal structures. Also, there has been no disbonding of those patches. The application of such repairs to thick sections or castings, on the other hand, has not proved to be as durable – and should not be expected to be so. However, the periodic replacement of a cheap patch as it disbonds is preferable to scrapping the part whenever no permanent repair is possible. Likewise, for military aircraft, temporary bonded patches can permit aircraft to remain in service until there is an opportune time for a permanent repair. Commercial aircraft operators tend to use temporary repairs only for ferry flights to better repair facilities.

In some ways, the topic of life prediction can be divorced from the design of repairs. However, it is quite common with metal aircraft structure to justify a mechanical repair by performing a static strength analysis alone and omitting a fatigue analysis because the remaining service life is considerably less than for new structure. The preceding discussion is intended to show that, for adhesive-bonded repairs which have been designed and performed properly, there is no reason to anticipate any reduction in service life due to the repair. The absence of loaded fastener holes and making the adhesive stronger than the surrounding structure means that the life is determined by that surrounding structure and not by the repair.

Much of the current research on wearout of adhesive bonds will contribute to the body of information on which good designs are based. Already, it is clear that fatigue or creep failures occur only under artificially severe test conditions, not representative of well-designed structures and that, in itself, is very reassuring for those promoting adhesive-bonded structures. The need for testing under such extremes arises from the lack of any failure to observe under realistic conditions. The author is confident that as such investigation is pursued, evidence will be accumulated to support the approaches recommended here. The concern is not primarily with the research itself but with the use that regulatory agencies are most likely to make of it. The current research should produce a better definition of the joint geometries and load conditions which should be excluded from the structural bonded joint designs. However, some of this research is being forced to conform to the framework of the analysis methods being developed for the damage tolerance and wearout of metal structure components by crack growth, and the author fears that similar analyses will be mandated for the adhesive bonds, without considering that the conditions are not at all analogous. Only if the adhesive were the critical element in the load path would such calculations be necessary, and in such a case, all of the established benefits from adhesive bonding would be lost. The author is trying to prevent the development of analysis requirements which are applicable only to inferior structures.

3.6 Surface preparation for adhesive-bonded repair of metal structure

The preceding statements that adhesive-bonded repairs should be the same as bonded joints in the original structure are echoed in Boeing's approach to surface preparation for bonding of aluminium alloys. Not only did Boeing pioneer the use of the phosphoric acid anodize surface treatment, which was also used on the PABST program at Douglas, but Boeing also developed nontank anodize techniques to permit durable repairs to be made on existing structure [10]. The disbonding and subsequent corrosion shown in Figure 3.10 make it apparent that there is no point in repairing bonded structure unless a good stable oxide surface is prepared by phosphoric- or chromic-acid anodize and protected by a good corrosion-resistant primer such as BR-127 or Redux resin. Details of the surface preparation techniques and environmental testing of bonded joints may be found elsewhere [11]. Recent developments have established that the same two anodize treatments are also superior to earlier surface treatments used on titanium alloys. Since the phosphate coating on automobile bodies has long been instrumental in improving the durability of the paint finish, it would not be surprising to discover one day that similar anodizing has also been shown to be effective for steels. It has also been said that phosphoric acid anodize on aluminium has at last solved the key problem of weld bonding – the conflicting surface requirements for spot welding and adhesive bonding. The PABST program has already brought tremendous advances in adhesive bonding of aircraft structure. But one of its most important legacies is the renewed interest in all aspects of adhesive bonding, not just in bonding aluminium alloys with ductile adhesives for use on subsonic transports. There is a quiet revolution in the surface treatments for other metal alloys.

Whereas chromic-acid anodizing leaves aluminium alloy panels more resistant to corrosion outside the bonded areas than they were before the treatment, the converse is true for phosphoric-acid anodizing. Indeed, the latter surface is so much more prone to corrosion that it must be protected by a 100 per cent coverage of corrosion-resistant primer and the usual layers of paint. Such protection was shown to be straightforward and effective during testing on the PABST program. The cure for the corrosion problem facilitates the repair of bonded structure. The in-service repair of a bonded panel fabricated with the phosphoric-acid anodize surface treatment is simplified because all of the surfaces have already been anodized and primed. So the field repair can be made without nontank anodize equipment. The detail parts can be anodized and primed off the aircraft in a conventional tank farm. The damaged area on the structure can be cut away, the edges treated with alodine, and the paint over the primer removed by chemical stripper. This is a powerful reason for using a separate primer during initial fabrication, even if it were not needed for other reasons. It also indicates that all corrosion-inhibiting primers must not be removed by the chemicals used to remove the other coatings.

Two cautions about adhesive-bonded repairs should be mentioned. First, solvents can be used effectively to remove gross contamination from an area to be repaired. They can also probably be used safely later in the cleaning cycle in a vapor degrease bath. But they must not be used as a solvent wipe on a rag to remove dust at the end of the cleaning cycle. That operation takes any residual contamination in the solvent or locally on the surfaces and spreads it around as a thin, uniform film over the entire surface to be bonded. Any specks of dust on the surface should be removed by a Scotchbrite pad or simply left there – after all, there are fillers in adhesive anyway, so a little more will not hurt. Perhaps the most persuasive argument against the use of the so-called solvent wipe was reported in a series of comparative surface preparation tests; a straight machined aluminium surface, still contaminated by drops of cutting oil, produced stronger bonds than were achieved by cleaning aluminium sheet with solvent.

The other caution concerns the use of room-temperature-cured paste adhesives to improve the fatigue life of riveted repairs. Such a scheme can be made to work, as shown by splices on the Fokker F-28 fuselages. But that application relies on the environmentally resistant surface preparation achieved by chromic-acid anodize and the coating of that splice area with Redux resin (as a primer) at the same time as the remainder of the bonded assembly was hot-bonded. It also relies on the fasteners to restrict the adhesive creep from accumulating under the sustained cabin pressure loads. Many in-service bond repairs made at room temperature, with or without rivets, have a typical life of about one week before the adhesive stops acting as a shear transfer medium and starts acting as a wet sponge to accelerate the corrosion of the faying surfaces. If the surfaces to be bonded cannot be prepared properly, the most effective repair is by rivets alone, with a good rubber-based faying surface sealant which does not need any special surface preparation to be effective. In thin sheet metal structure like control surfaces, which are subjected to loads of only short duration, the sealants tend to behave like adhesives anyway. It is only under sustained loads like cabin pressure that sealants do not contribute significantly to load transfer. Sealants are particularly effective for transferring load and preventing fatigue damage due to high-frequency, acoustically induced vibrations and probably could replace structural adhesives in many more situations than they have been used in to date.

3.7 Conclusions

This paper has shown that there is far more to consider about adhesive bonded structural repairs than merely designing a joint to carry some specified design load.

Indeed, for most metal-bonded structures, the design of such repairs has been reduced to finding the appropriate overlap in a table or multiplying the thickness by 30 or 80 for double- and single-shear, respectively. The most challenging bonded repairs in metal structure are associated with the stiffeners rather than the skin splices. The similarities between repair joints and original joints have been emphasized. In particular, it is important to note that any bonded joint should always be stronger than the members it is joining. It has been suggested that many small disbonds are better left alone than repaired. By and large, existing repair criteria are unrealistically stringent from the structural viewpoint. The whole subject of repair criteria needs a reassessment from an economic viewpoint because minimization of cost in series production would undoubtedly lead to higher quality bonded structures. Just as with bonding in original production, repairs must be made using the correct surface preparations, primers, adhesives, and cure cycles. Although it is sometimes difficult to effect adhesive-bonded structural repairs and mechanical fastening may be more practical, it is important to note that when adhesive bonding is appropriate, it is possible to restore the structure to 100 per cent of its original strength and service life.

References

[1] Thrall, Jr., E.W., *et al.*, Primary adhesively bonded structure technology (PABST): Design handbook for adhesive bonding. USAF Technical Rept. AFFDL-TR-79-3119 (November 1979).

[2] Hart-Smith, L.J., Adhesive bonding of aircraft primary structures. Douglas Aircraft Company Paper 6979, presented to SAE Aerospace Congress and Exposition, Los Angeles, California (October 1980).

[3] Hart-Smith, L.J., Structural details of adhesive-bonded joints for pressurized aircraft fuselages. Douglas Aircraft Company, Technical Reprt. MDC-J8858 (December 1980).

[4] Hart-Smith, L.J., Adhesively-bonded double-lap joints. NASA Langley Research Centre Rept. NASA CR-112235 (January 1973).

[5] Hart-Smith, L.J., Design and analysis of bonded repairs for fibrous composite aircraft structures. Douglas Aircraft Company Paper 7133, presented to International Workshop of Defence Applications of Advanced Repair Technology for Metal and Composite Structures, Naval Research Laboratory, Washington, D.C. (July 1981).

[6] Hart-Smith, L.J., Effects of adhesive layer edge thickness on strength of adhesive-bonded joints. Douglas Aircraft Technical Rept. MDC J4675 (May 1981).

[7] Hart-Smith, L.J., Differences between adhesive behaviour in test coupons and structural joints. Douglas Aircraft Company Paper 7066, presented to ASTM Adhesives Committee D-14 Meeting, Phoenix, Arizona (March 1981).

[8] Hart-Smith, L.J., Nonlinear analysis of bonded/bolted joints. To be published as Part 4 of Design Methodology for bonded-bolted composite joints, analysis derivations and illustrative solutions. Douglas Aircraft Company, for USAF Flight Dynamics Laboratory.

[9] Hart-Smith, L.J., Adhesive bond stresses and strains at discontinuities and cracks in bonded structure. J. of Engineering Materials and Technology, Vol. 100, pp. 16–24 (January 1978).

[10] Locke, M.C., Horton, R.E. and McCarty, J.E., Anodize optimization and adhesive evaluation for repair applications. USAF Technical Rept. AFML-TR-78-104 (July 1978).

[11] Shannon, R.W., *et al.*, Primary adhesively bonded structure technology (PABST): General material property data. USAF Technical Rept. AFFDL-TR-77-107 (September 1978).

R. JONES

4

Crack patching: design aspects

4.1 Introduction

This chapter describes two complementary approaches to the analysis and design of bonded repairs. First, an approach based on the finite element method is presented and illustrated by an application to an actual repair (Sections 4.2 and 4.3). Next, an analytical approach is described which leads to explicit estimates for the design parameters for an idealized case and for repairs to composite sheets (Section 4.5). In Section 4.6, the results of a three-dimensional finite element analysis of an edge-cracked plate, repaired by a bonded semi-circular reinforcing patch, are presented and compared with the results of the two-dimensional analysis of Sections 4.2 and 4.5 as well as with experimental measurements on a test specimen. Finally, two design studies of repairs to surface cracks, and cracked holes, are described, in Section 4.7 and 4.8 respectively.

There are several methods available for patch design when cracking occurs in thin sheet, i.e. typical sheet thickness of 3 mm [1–8]. In these methods, the skin is considered to be in a state of plane stress and bending effects are ignored. However, the induced bending effects which result from patching may be accounted for in each method, see [4]. These methods are categorized as follows:

(a) Finite element approaches [5, 6].
(b) Analytical formulae based on certain simplifying assumptions [7, 8].
(c) Other approaches [1–4].

Each method yields similar results, although the finite element approach was the first to be developed and is the approach used to design complex repair schemes, e.g. for the Mirage III and the F111 console truss [9, 10].

The analytical formulae are particularly easy to use and provide a first estimate for the patch design. In some cases this first estimate is sufficient. However, there are situations in which the repair is critical and a long life is required. In these cases a full finite element analysis is necessary, e.g. [9, 10]. As a result the finite element approach will be discussed first and illustrated by considering the design of a repair to the lower wing skin of a Mirage III aircraft. This design study highlights the critical design parameters.

4.2 The finite element formulation

When analyzing bonded repairs to cracked metallic sheets it is first necessary to develop a realistic mathematical model for the behaviour of the adhesive layer bonding the patch to the sheet. Under inplane or transverse loading, shear stresses are developed in the adhesive. If we define the x and the y axes to lie in the plane of the sheet with the z axis in the thickness direction then these shear-stresses can be expressed in terms of the displacements in the sheet and the patch, viz:

$$\tau_{sx} = K_1(u_R - u_P) + K_3(v_R - v_P) + K_4\frac{\partial w}{\partial x} + K_6\frac{\partial w}{\partial y}$$

$$\tau_{sy} = K_3(u_R - u_P) + K_2(v_R - v_P) + K_6\frac{\partial w}{\partial x} + K_5\frac{\partial w}{\partial y} \tag{4.1}$$

Here τ_{sx} and τ_{sy} are the values of the shear stresses in the adhesive K_1, K_2, K_3, K_4, K_5 and K_6 are spring constants whose values depend on the material properties and thickness of the adhesive, skin and composite patch. The terms u_R, v_R and u_P, v_P are the displacements at the midsurface of the patch and the skin respectively while w is the vertical deflection. It is often a reasonable assumption that for the composite patch, which from here on will be considered to be unidirectional with the fibres perpendicular to the crack, $G_{13} = G_{23} = G_{12}(G_R)$, i.e. the interlaminar shear moduli are equal. This assumption is unnecessary and the full form for the K_i's is contained in [5, 6]. However, it dramatically simplifies the analysis and has little effect on any quantities of interest. With these assumptions we obtain, in the case of a patch on one side of the skin

$$K_1 = K_2 = 1/(t_A/G_A + 3t_R/8G_R + 3t_R/8G_P)$$

$$K_3 = K_6 = 0$$

$$K_4 = K_5 = (t_A + t_R/2 + t_P/2)K_1 \tag{4.2}$$

where t_A, t_R and t_P are the thicknesses of the adhesive, patch (composite overlay) and skin respectively and G_A, G_R and G_P are the shear modulii of the adhesive, patch and skin.

These formulae were first obtained in [5, 6] and make full allowance for the shear deformation which occurs throughout the composite patch and skin. With this approach, the u and v displacements through the patch, adhesive and skin are given by

$$u = u_R + \tau_{sx}(z(t_P + t_A + t_R - z/2) - (t_P + t_A + t_R/2)(t_P/2 + t_A/2 + 3t_R/4))/G_R t_R \text{ in the patch}$$

$$u = (u_R - 3\tau_{sx}t_R/8G_R)(z - t_P)/t_A + (u_P + 3\tau_{sx}t_P/8G_P)(t_P + t_A - z)/t_A \text{ in the adhesive} \tag{4.3}$$

$$u = u_P + (z^2 - t_P^2/4)\tau_{sx}/2G_P t_P \text{ in the sheet}$$

and a similar expression for v.

If a patch is placed on both sides of the skin, then the term $3t_P/8G_P$ appearing in the above equations is replaced by $t_P/4G_P$. These assumptions result in a shear stress profile which is piece wise linear.

4.2.1 Element stiffness matrix

Having obtained the nature of the stress field in the adhesive we can now derive the stiffness matrix for the adhesive layer. When there is no bending the sheet is assumed to be in a state of plane stress and it is usually modelled by isoparametric membrane elements while the patch is modelled by isoparametric membrane elements with an orthotropic stress strain law. The adhesive is also assumed to only carry the shear stresses τ_{xz} and τ_{yz}. As a result the total strain energy of an element of the repaired structure is

$$V = \tfrac{1}{2}\boldsymbol{\delta} K_P \boldsymbol{\delta}^T + \tfrac{1}{2}\boldsymbol{\delta} K_R \boldsymbol{\delta}^T + \frac{1}{2}\iiint(\tau_{xz}\gamma_{xz} + \tau_{yz}\gamma_{yz})\mathrm{d}x\mathrm{d}y\mathrm{d}z \tag{4.4}$$

where K_P and K_R are the stiffness matrices of the skin and the patch respectively. The last term in this expression is the contribution due to shear deformation. To be completely accurate the z integration should be over the total thickness of the skin, adhesive and patch whilst the x, y, integration is over the surface area of the element.

We first define a vector **f** such that

$$\mathbf{f} = (u_R, v_R, u_P, v_P) = (N\boldsymbol{\delta})^T \tag{4.5}$$

where as explained in the book by Zenkiewicz [11] on finite elements the components of N are generalised functions of position and

$$\boldsymbol{\delta}^T = (\delta_1^T, \delta_2^T \ldots \delta_m^T). \tag{4.6}$$

Here the element is considered to be an arbitrary shape with m nodes and

$$\delta_i^T = (u_{Ri}, v_{Ri}, u_{Pi}, v_{Pi}) \tag{4.7}$$

The strain vector may be expressed as

$$\gamma^T = (\gamma_{xz}, \gamma_{yz}) = (D\boldsymbol{\tau})^T \tag{4.8}$$

where the matrix D is a function of z and where

$$\boldsymbol{\tau}^T = (\tau_{sx}, \tau_{sy}) \tag{4.9}$$

However, making use of equations (4.1), (4.5), we find that τ can be written as

$$\boldsymbol{\tau} = AN\boldsymbol{\delta} \tag{4.10}$$

where

$$A = \begin{bmatrix} 1, & 0, & -1, & 0 \\ 0, & 1, & 0, & -1 \end{bmatrix}(t_A/G_A + 3t_P/8G_P + 3t_R/8G_R). \tag{4.11}$$

As mentioned before the assumed deformation profile results in a quasilinear shear stress. The shear stresses τ_{xz} and τ_{yz} vanish at the top of the patch increase linearly to the values of τ_{sx} and τ_{sy} respectively in the adhesive and decay linearly to zero at the free surface of the skin (or plane of symmetry in the case of a patch on both sides of the skin). This results in the matrix D having the form

$$\begin{aligned} D &= \frac{f(z)}{G_P}\begin{bmatrix}1 & 0\\ 0 & 1\end{bmatrix} \text{ in the skin, i.e. } 0 \leqq z \leqq t_P \\ &= \frac{f(z)}{G_A}\begin{bmatrix}1 & 0\\ 0 & 1\end{bmatrix} \text{ in the adhesive } t_P \leqq z \leqq t_P + t_A \\ &= \frac{f(z)}{G_R}\begin{bmatrix}1 & 0\\ 0 & 1\end{bmatrix} \text{ in the patch, } t_P + t_A \leqq z \leqq t_P + t_A + t_R \end{aligned} \tag{4.12}$$

with the term t_P being replaced by $t_P/2$ in the case of a patch on both sides of the skin. Here

$$\begin{aligned} f(z) &= z/t_P \text{ in the skin } z \leqq t_P \\ &= 1 \quad \text{in the adhesive } t_P \leqq z \leqq t_P + t_A \\ &= (t_P + t_A + t_R - z)/t_R \text{ in the patch } t_P + t_A \leqq z \leqq t_P + t_A + t_R \end{aligned} \tag{4.13}$$

Making use of the expression for τ_{xz}, γ_{xz}, etc we now find that

$$\begin{aligned} &\frac{1}{2}\iiint(\tau_{xz}\gamma_{xz} + \tau_{yz}\gamma_{yz})\mathrm{d}V \\ &= \frac{1}{2}\iint (AN\delta)^T\left(\int f(z)D\mathrm{d}z\right)AN\delta \,\mathrm{d}x\mathrm{d}y \end{aligned} \tag{4.14}$$

As a result, we find that the stiffness matrix K^e for the adhesive layer, plus the shear coupling in the skin and patch, is given by

$$\begin{aligned} K^e &= \iint(AN)^T\left(\int f(z)D\mathrm{d}z\right)AN \,\mathrm{d}x\mathrm{d}y \\ &= (t_A/G_A + t_R/3G_R + t_P/3G_P)\iint(AN)^T\begin{bmatrix}1 & 0\\ 0 & 1\end{bmatrix}AN \,\mathrm{d}x\mathrm{d}y \end{aligned} \tag{4.15}$$

For a patch on both sides of the skin, the stiffness matrix for both layers of adhesive and the shear coupling in the skin and patch is

$$K^e = 2(t_A/G_A + t_R/3G_R + t_P/6G_P)\iint(AN)^T\begin{bmatrix}1 & 0\\ 0 & 1\end{bmatrix}AN \,\mathrm{d}x\mathrm{d}y \tag{4.16}$$

A more general expression for K^e including bending effects is given in [14].

Having thus obtained the stiffness matrix for adhesive layer it is now possible to analyse complex repair schemes.

To illustrate the versatility of this approach we will consider the boron fibre repair to fatigue cracks in the lower wing skin of Mirage III aircraft in service with the RAAF. This problem highlights the recommended approach to designing bonded repairs to cracked metallic wing skins.

4.3 Repair of cracks in Mirage III lower wing skin – a design study

In the late 1970's the Royal Australian Airforce tasked the Aeronautical Research Laboratories to develop a boron fibre patch for cracks in the lower wing skin of a number of Mirage III aircraft in service with the RAAF. These cracks were predominantly found at an angle of 45 degrees to the main spar. To investigate the feasibility to a b/ep repair a design study was undertaken into the repair of a crack whose tips were 111 mm apart and which lay at 45 degrees to the spar (see Figure 4.1). This crack was a model of the longest crack which had been found in service.

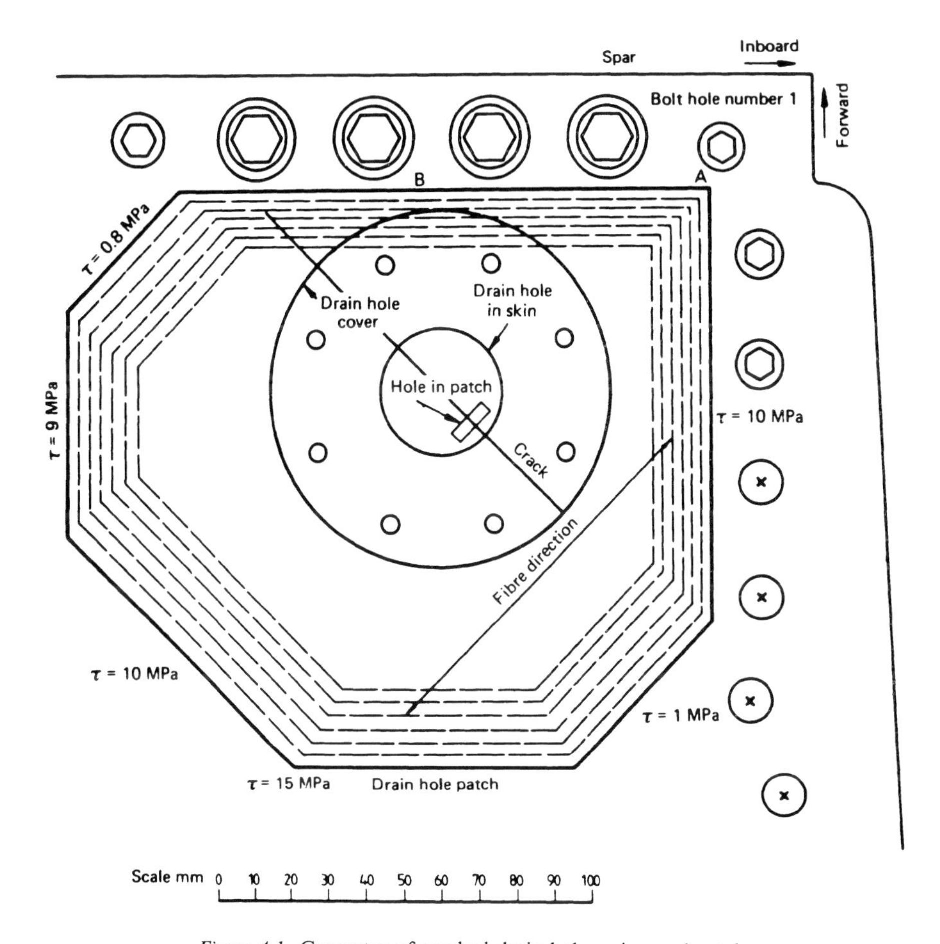

Figure 4.1. Geometry of cracked drain hole region and patch.

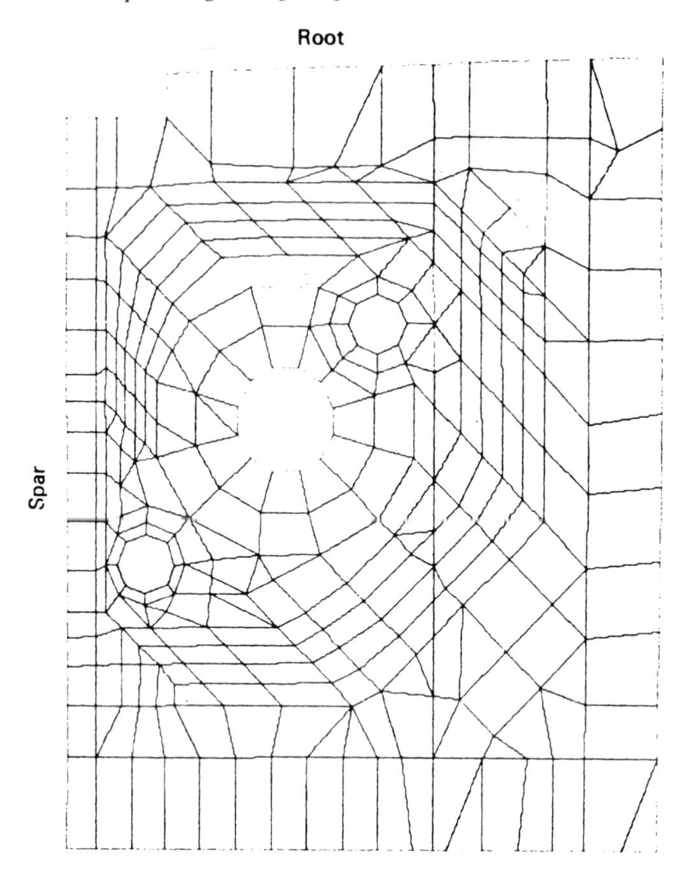

Figure 4.2. Finite element mesh for drain hole region.

As a first step in the design of the repair a study of the cracked, but unpatched, region was undertaken. A detailed finite element model of the area surrounding the drain hole region was developed, see Figure 4.2. The loads applied to this model were obtained directly from the stress distribution which resulted from a previous finite element model of this region and which correspond to a 7.5 g load case.

The study gave the values of the stress intensity factors to be $k_1 = 72\,MPa\sqrt{m}$ and $k_2 = 3.3\,MPa\sqrt{m}$ at the tip closest the spar and $k_1 = 68\,MPa\sqrt{m}$, $k_2 = 0.5\,MPa\sqrt{m}$ at the tip closest to the root rib. These values are consistent with a fractographic examination of the crack which showed that the crack was essentially growing as a mode 1 fracture and that of the two crack tips the tip closest to the spar was growing the faster. Indeed the tip closest to the spar was found to be very close to final failure.

Having thus obtained a reasonable model for the unpatched crack we add to this a finite element representation of the repair. Six b/ep patch configurations were considered, each with the same plan form, see Figure 4.1. Each patch was modelled using approximately 380 of the "bonded" elements described in Section 4.2.

All of the six patches considered were unidirectional laminates and were internally stepped, i.e., with the longest ply on the outside. The fibre direction is at ninety degrees to the crack.

Initially it was uncertain if carrying the fibres over the drain hole was necessary, or how frequently the drain hole was used in service. As a result, in three of the patches considered a hole was left so as not to interfere with the draining of the wing. In the other three patches varying amounts of the hole were covered. In one case one third of the total area of the hole was covered while in the other two cases virtually all of the hole was covered.

For each of the patches, the maximum shear stresses in the adhesive bonding the patch to the wing skin occurred at points *A*, *B*, *C* and *D* (see Figure 4.1). The maximum stresses in the fibres occurred at point *D* for the patches with a hole in the patch, and in the fibres over the hole in the patches with the hole partially covered. The values for these stresses, along with the percentage reduction of the stress intensity factor K_{I} at each tip, achieved by each patch are shown in Table 4.1. Here $K_{\mathrm{I}u}$ and $K_{\mathrm{I}p}$ are the values of the stress intensity factors before and after patching respectively.

We see that all of the six patches achieve a reduction in the stress intensity factor K_{I} of at least 91%. Consequently they would all significantly retard growth. Similarly for all of the patches, the fibre strains are below the maximum working levels of 0.005, which corresponds to a stress of 1 GPa, although of the six patch

Table 4.1. Effect of drain hole patch; 7.5 g load case.

Patch number	1	2	3	4	5	6
Maximum patch[2] Thickness (mm)	0.762	0.762	0.776	0.889	0.889	0.889
Adhesive thickness (mm)	0.102	0.102	0.102	0.203	0.203	0.203
Thickness of first layer (mm)	0.127	0.254	0.127	0.254	0.254	0.127
Covering of drain	open	open	1/3 covered	open	substantially covered	substantially covered
Adhesive shear stress (MPa) at points						
A	29	43	29	31	26	18
B	55	79	55	58	42	30
C	181	179	164	120	63	64
D	153	153	131	98	50	51
Maximum fibre stress (MPa)	953	930	760	911	450	455
Reduction in stress intensity factor K_{I} i.e. $1 - k_{1p}/k_{1a}$, at:						
1. Spar tip	91%	91%	91%	91%	92%	91%
2. Root rib tip	99%	99%	99%	99%	99%	99%

numbers 5 and 6 have by far the greatest factors of safety. As a result the patch design was finally chosen primarily on the basis of the magnitude of the shear stresses developed in the adhesive. On this basis patch numbers 1, 2, 3 and 4 were rejected. The two remaining patches are patch numbers 5 and 6. Of the two, Table 4.1 shows that the adhesive shear stresses along the edges of the patch, are substantially higher for patch number 5 than for patch number 6, although both are below the threshold value for fatigue damage. As a result, patch 6 is much less likely to suffer fatigue damage to the adhesive bond. Hence, patch number 6 was adopted as the final repair.

Consulting Table 4.1, we see that at locations C and D in patch 6 the shear stress in the adhesive is sufficiently high so as to cause concern over the possibility of fatigue damage occurring in the adhesive. However, these high values occur in the interior of the patch at the intersection of the crack with the drain hole, and are very localized. As a result any damage which does occur should not spread and, as shown in [5, 6], should have virtually no effect on the stress intensity factors at the crack tips or on the fibre stresses.

Let us now summarize the criteria which have been used to finalize the patch design.

1. The peak fibre stresses must be less than the maximum permissible tensile stength. For boron epoxy this is approximately 1 GPa.
2. The peak adhesive stress must be less than its fatigue threshold value. For the AF126 epoxy nitrile adhesive used, this is approximately 40 MPa.
3. A significant reduction in the stress intensity factor k_{1p} must be achieved. Wherever possible, it is desirable to reduce k_{1p} to below the fatigue threshold limit of the wing skin.

Let us now consider the effect that the difference between the coefficients of thermal expansion of the boron patch and aluminium alloy wing skin has on the residual stresses left in the skin after the patch has been applied. Patching the skin involved heating the area to be repaired to approximately 120°C. The patch which has also been heated to 120°C, is then attached and the patched structure is allowed to cool to ambient temperature. It is during this cooling phase that the difference in the coefficients of expansion between the patch, $\alpha_1 = 4.5 \times 10^{-6}$ per °C and $\alpha_2 = 20 \times 10^{-6}$ per °C, and the wing skin, $\alpha = 23 \times 10^{-6}$ per °C, causes a residual thermal stress to be left in the structure. (Note that α_1 is the coefficient of expansion in the fibre direction and α_2 is that perpendicular to the fibre direction).

In order to analyse this phenomenon the shear modulus of the adhesive was taken as zero at 120°C and was assumed to increase linearly, as the temperature decreased, to a value of 0.965 MPa at ambient temperature i.e. 40°C. At each step decrease in temperature the adhesive layer was modelled using the finite element method described above. As a result of this analysis, it was found that when the skin was assumed to be restrained from in plane movement by the spar and root rib attachments, the mean residual stress left in the skin under the patch was a tensile stress of 8.7 MPa. This stress should not significantly alter fatigue behaviour of the patched panels.

A more detailed discussion of the effects of thermal mismatch is given in [10, 12] as well as in Chapter 6 of this book.

4.4 Neutral axis offset effects

Until now we have ignored bending effects. However, even if the wing skin is in a state of plane stress the location of the neutral axis of the patch-adhesive-skin section will differ from the neutral axis of the wing skin itself. Hence forces applied to the skin will result in out of plane bending which will reduce the efficiency of the repair.

A method, developed at Northrop [3], may be used to account for this out of plane bending. In this method the apparent stress intensity factor K_p^* at the mid surface of the sheet is given by

$$K_p^* = (1 + BC)K_p \tag{4.17}$$

where BC is a bending correction factor. Here

$$BC = y_{\max} a(1 - K_P/K_S)t_P(t_P + t_R)/I \tag{4.18}$$

where K_S is the value of the stress intensity factor before patching, t_P and t_R are the thicknesses of the sheet and patch respectively, $y_{\max}$ is the distance of the lower unpatched surface of the plate from the neutral axis of the section (i.e. sheet plus patch), I is the moment of inertia of the section and a is the crack half length.

The value of K_P^* can be related, see [14], to J, the energy release rate for self similar crack growth, in the usual way viz:

$$J = K_P^{*2}/E. \tag{4.19}$$

However, since the growth is non self-similar with the maximum growth occurring at the lower free surface it is best not to use K_P^* or J but to design on the basis of the maximum stress intensity factor $K_P^{\max}$ which is given by

$$K_P^{\max} = (1 + 2BC)K_P \tag{4.20}$$

In the previous sections we have attempted to outline the analytical and numerical tools necessary to design a repair. Indeed these procedures can be used even in the most complex and critical situations.

Having identified the critical parameters for patch design let us now turn our attention to simple analytical methods which enable an initial design to be made without the need for complex computational analyses.

4.5 Initial design procedures

Following the development of the finite element approach, an approximate closed form solution was obtained for the repair of a centre notch panel. This solution was based on the assumption that bending does not occur and that for a sufficiently long crack the central region is behaving like a bonded overlap joint, see [7, 8].

The important results modified so as to allow for shear deformation, arising out of this approximate solution may be summarized as follows.

1. After patching the stress intensity factor asymptotes to a constant value K_∞ as the crack length increases. This value may be approximated by

$$K_\infty = \sigma_0\sqrt{\pi\lambda} \qquad (4.21)$$

where

$$\pi\lambda = [\beta^{-1}E_P t_P(1 + t_P E_P/E_R t_R)]^{1/2} \qquad (4.22)$$

$$\sigma_0 = E_P t_P \sigma/(E_R t_R + E_P t_P) \qquad (4.23)$$

and

$$\beta = (t_A/G_A + t_R/3G_R + t_P/3G_P)/(t_A/G_A + 3t_R/8G_R + 3t_P/8G_P)^2. \qquad (4.24)$$

2. The peak adhesive stress τ_{max} over the crack asymptotes to the value

$$\tau_{max} = K_a \sigma_0 \pi\lambda/2E_P \qquad (4.25)$$

where

$$K_a = 1/(t_A/G_A + 3t_R/G_R + 3t_P/8G_P). \qquad (4.26)$$

3. The peak fibre stress σ_f in thc patch asymptotes to the value

$$\sigma_f = \sigma t_P(1 + P)/t_R \qquad (4.27)$$

where

$$P = (\varrho - 1)/D[2\varrho m + 1 - \nu^2 - \nu(1 - \nu)\varrho] \qquad (4.28)$$

$$D = (2\varrho/m + 1)(2\varrho m + 1) - (\nu + (1 - \nu)\varrho)^2 \qquad (4.29)$$

$$\varrho = E_I/E_P \qquad (4.30)$$

$$E_I = E_P + E_R t_R/t_P \qquad (4.31)$$

$$m = a/b \text{ (the aspect ratio of the repair)} \qquad (4.32)$$

and where ν is the Poisson's ratio of the skin. Here t_P is one half of the sheet thickness.

When only one side of the sheet is patched, the limiting stress intensity factor K_{max} is given by

$$K_{max} = (1 + 2BC)K_\infty \qquad (4.33)$$

where BC is the bending correction factor given in the previous section.

These formulae provide fairly accurate estimates of the critical design parameters and are very useful in obtaining an initial design.

4.5.1 Bonded repairs to fibre composite plates

Up till now we have been primarily concerned with repairs to cracked metallic components. Yet the analytical and numerical tools developed are also directly applicable to the repair of damaged fibre composite plates. Indeed the same arguments as used in [7, 8] can also be used to show that the energy release rate $\mathscr{G}$ ($= K^2/E$ for an isotropic material) for a cracked composite plate repaired by a bonded patch also goes to a limiting value, viz; $\mathscr{G}_\infty$, as the crack length increases. An estimate for this value is

$$\mathscr{G}_\infty = \sigma_0^2 \pi\lambda / E_{11P} \quad (4.34)$$

where

$$\sigma_0 = \sigma E_{11P} t_P / (t_P E_{11P} + t_R E_{11R}) \quad (4.35)$$

$$\pi\lambda = [P^{-1} E_{11P} t_P (1 + E_{11P} t_P / E_{11R} t_R)]^{1/2} \quad (4.36)$$

and where

$$P = \frac{(t_A/G_A + t_R/3G_R + t_P/3G_P)}{(t_A/G_A + 3t_R/8G_R + 3t_P/8G_P)^2}. \quad (4.37)$$

Here the suffix 1 refers to the direction perpendicular to the crack and for the sake of simplicity, we have assumed that the interlaminar shear moduli in the plate are equal and take the value G_P, and that the interlaminar shear moduli in the reinforcement are equal and have the value G_R.

When the interlaminar shear moduli differ significantly, a more complex analysis such as that presented in [14] may be required. Indeed the formulation of the adhesive stiffness matrix given earlier is still valid provided that G_P is taken to be the interlaminar shear moduli of the composite sheet.

4.6 Comparison with experimental and 3-D results

In order to evaluate the accuracy of the analytical tools developed in the previous sections, let us consider a rectangular sheet of aluminium alloy with dimensions 150 mm × 320 mm × 3.15 mm containing an edge crack, of variable length, which lies along the centre line of the sheet (see Figure 4.3).

The crack is patched with a uni-directional boron epoxy laminate 0.889 mm thick, the direction of the fibres being perpendicular to the crack. The adhesive is 0.1651 mm thick, has a Young's modulus $E_A = 1.89$ GPa and a shear modulus $G_A = 0.7$ GPa. The moduli of the boron epoxy laminate are taken as

$$E_{11} = 208.3\,\text{GPa}, \quad E_{22} = E_{33} = 25.4\,\text{GPa}$$

$$G_{13} = G_{23} = G_{12} = 7.24\,\text{GPa}, \quad \nu_{13} = \nu_{12} = 0.183$$

$$\nu_{23} = 0.1667$$

whilst the Young's modulus and Poisson's ratio of the aluminium are 72.86 GPa and 0.3 respectively. We first analyse this problem, making use of symmetry, using a fully three-dimensional finite element model. The aluminium sheet is modelled by forty-one twenty-noded isoparametric bricks and thirteen of the fifteen-noded isoparametric elements whilst the composite patch is represented by twenty-one of

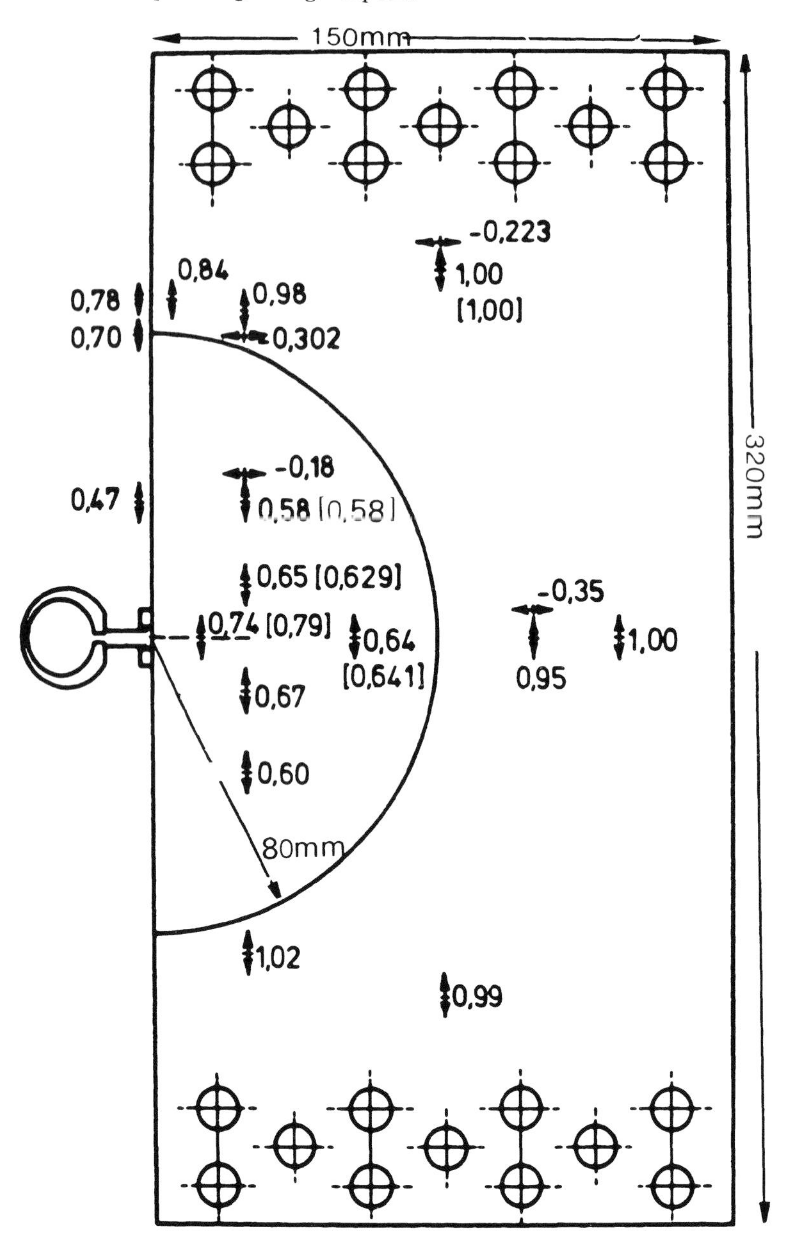

Figure 4.3. Ratio of measured strains to the far field strain, finite element values in brackets.

the twenty-noded isoparametric bricks and thirteen of the fifteen-noded isoparametric elements.

The elements at the crack tip are triangular in planform and have the midpoint nodes moved to the quarter points in order to simulate the $r^{-1/2}$ singularity at the crack tip.

To avoid problems with numerical ill conditioning and the use of elements with large aspect ratios, reduced integration [11] must be used whenever a full three dimensional analysis is undertaken. In addition the formulation of the stiffness matrices and the solution must be done using double precision on 32 bit machines.

The values of the stress intensity factors at various points along the crack front are shown in Table 4.2 for various crack lengths for the case of a uniform applied tensile stress of 137.9 MPa and for which bending is not allowed.

Here we see that the maximum value of the stress intensity factor does indeed asymptote to a constant value. However, in this analysis, the convergence is not from below as postulated in [7] and the smaller crack lengths have larger stress intensity factors. Consequently, it is possible that after patching short cracks will grow faster than long cracks.

It is also very important to note that there is a significant variation in the fibre stresses across the thickness of the repair. At first glance, this would eliminate the possibility of using the analytical formulae to predict the peak fibre strain. However, in practice, failure of a composite is determined by the average stress law, viz:

$$\frac{1}{a_0}\int \sigma_y \mathrm{d}z = \sigma_u. \tag{4.38}$$

Here the integral is taken over the patch thickness, a_0 is a characteristic length, σ_u is the failure stress of the basic laminate and σ_y is the fibre stress. For boron and carbon epoxy laminates, a_0 is generally greater than the laminate thickness, so that this failure law can be replaced by

$$\frac{1}{t_R}\int \sigma_y \mathrm{d}z = \sigma_u. \tag{4.39}$$

The left hand side of equation (4.39) is the mean stress in the laminate and is obtainable using the simple formulae given in the previous section. We thus see that failure occurs when the mean stress exceeds the failure stress of the basic laminate.

At this stage we must emphasize that this mean stress level will not be experi-

Table 4.2.

	Crack length mm				
	6.35	12.7	19.3	25.4	50.8
K_{1p} $MPa\sqrt{m}$, at					
Adhesive/sheet interface	5.95	5.62	5.46	5.13	4.02
Mid surface	8.43	7.97	7.82	7.54	6.83
Free surface	7.76	7.50	7.53	7.54	7.54
Crack opening displacements (mm) at					
Adhesive/sheet interface	0.006	0.005	0.004	0.003	0.001
Mid surface	0.012	0.012	0.012	0.011	0.010
Free surface	0.013	0.015	0.016	0.017	0.017
Peak fibre stress MPa, at					
Top surface of patch	341	327	329	339	349
Mid surface	367	434	447	457	455
Bottom surface of patch	682	675	660	630	562

enced by the surface of the patch. Indeed, as we will shortly show, strain gauges on the patch will record only the surface strains which are considerably lower than the mean strain.

Let us now compare these results with those obtained experimentally for this repair configuration. Figure (4.4) shows a comparison of the numerically predicted surface strains with those measured strains on the surface of the patch at four locations. The clip gauge openings measured near the mouth of the crack are given in Table 4.3 as are those predicted numerically and those using the analytical approach.

This clearly shows that the finite element approach is capable of accurately modelling realistic repairs and that the analytical formulae provide a good estimate of the clip gauge opening.

The analytical model predicts an upper bound on the peak fibre stress of 580 MPa. Indeed this compares favourably with the peak mean fibre stress of 457 MPa obtained numerically.

The following conclusions may now be made:

1. The finite element approach is capable of providing detailed information concerning "real life" repair situations.

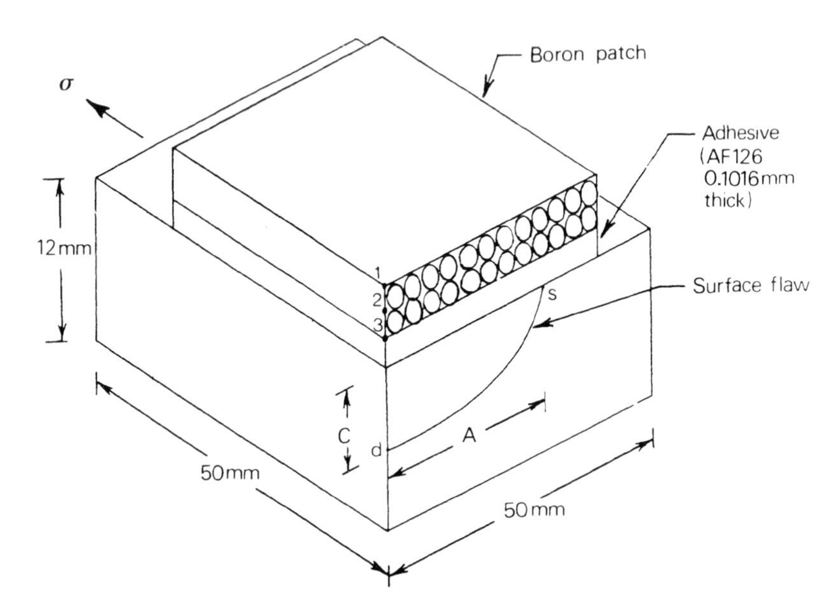

Figure 4.4. Repair of surface flaw (1/4 structure modelled).

Table 4.3. Clip gauge opening.

	Clip gauge opening (mm)
Predicted using F.E.M.	0.037 mm
Predicted analytically – (overlap joint analogy)	0.032 mm
Measured	0.041 mm

2. The analytical formulae provide good first estimates of the critical design parameters.

Further details concerning the effect of this patch on crack growth is given in Chapter 6.

4.7 Repair of semi-elliptical surface flaws

In recent years, a number of b/ep patches have been designed at ARL for surface flaws in thick sections, e.g. the repairs to the Macchi and Mirage III main landing wheels and the repair to the console truss in F111 aircraft (see [9]). In each case, the crack section was 12 mm thick.

In the case of the Mirage and Macchi landing wheels, the repairs are installed when the crack reaches a total length of 24 mm. In each case, the cracks were found to be nearly semi-elliptical in shape with a surface length of 24 mm and a maximum depth of 6 mm. In order to study the effect on such a crack, an investigation was undertaken on the repair of a similar semi-elliptical crack centrally located in a rectangular block of aluminium with dimensions as shown in Figure (4.4) (in this figure only one quarter of the structure is shown). The block was subjected to a uniform uniaxial stress, and the effect that various boron fibre patches had on the crack were calculated using a detailed three dimensional finite element analysis. Table 4.4 shows the calculated values of the stress intensity factors at point d, the point of deepest penetration, and s, the point at which the crack intersects the free surface. The fibre stresses σ_f are a maximum over the crack and vary through the thickness of the patch. These values are also shown in Table 4.4 along with the peak adhesive stresses over the crack. In this study, the adhesive was taken to be AF126, an epoxy nitrile, 0.1016 mm thick with a shear modulus of 0.7 GPa.

Table 4.5 shows the corresponding values of stress intensity factors, fibre stresses and adhesive stresses for the case when the the surface flaw is semi-circular, rather than semi-elliptical, with a surface length of 12 mm.

Using these Tables, it is possible to design a repair for surface flaws up to 24 mm in surface length. The major design considerations are as previously given, viz:

1. The maximum stress intensity factor should be as low as possible and preferably below the critical value for fatigue crack growth in the material.

Table 4.4. Semi-elliptical flaw A = 12 mm, C = 6 mm. Unpatched values at d, $K_1 = 12.45\ MPa\sqrt{m}$, at s, $K_1 = 12.5\ MPa\sqrt{m}$.

Number layers boron	Stress intensity factor K_1 at:		Fibre stress over crack σ_f/σ at points			Adhesive shear stress over crack τ/σ
	d	s	1	2	3	
5	8.33	4.396	3.481	4.256	5.293	0.451
10	7.222	3.463	1.918	2.572	3.670	0.353
15	6.662	2.979	1.271	1.810	2.907	0.304
20	6.365	2.709	0.874	1.364	2.488	0.277
25	6.211	2.588	0.587	1.073	2.244	0.262

Table 4.5. Semi-circular flaw A = C = 6 mm.

Number layers boron	Stress intensity factor K_I at:		Fibre stress over crack σ_f/σ at points:			Adhesive shear stress over crack τ/σ
	d	s	1	2	3	
5	6.882	5.412	3.333	3.966	4.811	0.402
10	6.232	4.552	2.113	2.672	3.615	0.337
15	5.842	4.035	1.541	2.005	2.972	0.300
20	5.605	3.712	1.153	1.576	2.582	0.276
25	5.465	3.509	0.852	1.276	2.337	0.261

2. The maximum adhesive stresses should be below the value at which fatigue damage accumulates in the adhesive. For AF126, which is an epoxy nitrile, this is 40 MPa.
3. The average stress through the thickness of the boron patch should not exceed 1000 MPa. This failure rule is a modification of the average stress failure criterion which is commonly used for composite materials.

There will be circumstances in which it is not possible to satisfy all three of these design rules using the unidirectional boron fibre patches which are considered here. In these cases, the possibility of using a purpose designed unbalanced patch such as first described by Jones [14] should be investigated. In cases where the fibre stress is the limiting design parameter hybrid graphite and boron epoxy patches, with the graphite epoxy plies placed nearest to the crack, should be considered.

4.7.1 Experimental verification

In order to verify the significant reductions in the stress intensity factors and hence in the rate of crack growth a series of fatigue tests were performed. The specimens were a 2024 T4 aluminium alloy, 11.1 mm thick, 108.3 mm wide and 304 mm long and contained a centrally located surface crack. The surface length of the crack was 37 mm and it was 6 mm deep. Five specimens were tested under a constant amplitude stress of 65 MPa and $R = 0.01$. A further six specimens were repaired with a ten ply thick boron fibre patch. In confirmation of our previous results, the unpatched specimens lasted an average of 22,450 cycles whilst the patched specimens lasted an average of 527,000 cycles. Even after failure there was no evidence of the patch debonding or of fibre breakage.

From Tables 4.4 and 4.5 we see that the peak fibre and adhesive stresses are relatively insensitive to the surface lengths of the crack. Consequently, we can estimate the peak values of these stresses in the specimens, viz:

$$\sigma_f = 3.67 \times 65\,\text{MPa} = 238\,\text{MPa}$$

and

$$\tau_{\max} = 0.353 \times 65\,\text{MPa} = 22.94$$

These values are considerably lower than their critical design values. Several specimens were also loaded at 130 MPa with no evidence of fibre breakage or patch debonding.

4.8 Repair of cracked holes

The fatigue life of cracked holes is of major interest to the aerospace industry and a great deal of effort has been spent on developing suitable repair schemes. Indeed during the full-scale fatigue testing of Mirage III fighter wings at the Swiss Federal Aircraft Factory (F + W), Switzerland, fatigue cracks were discovered at the innermost bolt holes along the rear flanges of the main spars. Subsequently, crack indications were confirmed at identical locations in wings of the Royal Australian Air Force Mirage III fleet [18]. As a consequence, several investigations were undertaken at the Aeronautical Research Laboratories (ARL) to explore methods for increasing the fatigue lives at critical sections of the spars [18–20].

The development of a life-enhancement scheme for this portion of the spar was complicated by the presence of two through-the-flange single-leg-anchor-nut(SLAN) rivet holes located in a chordwise direction and close to a critical bolt hole. Although the installation of interference-fit steel bushes at the bolt hole virtually inhibited crack initiation at the bolt hole [18, 19], and the adoption of a modified system for securing the SLAN obviated the need for through-the-flange rivets, a consequence was that the SLAN rivet holes then became the critical locations for fatigue crack initiation. Several alternatives including cold-expanding and the insertion of close-fit rivets were tested as part of the main life-enhancement investigation for the spar, but these did not provide any significant improvements in life for cracks which initiated at the rivet holes.

Adhesive bonding of close-fit rivets in the SLAN holes was then proposed as a method of improving fatigue life. This was based on the premise that the adhesive would result in better load transfer through the rivet (and in so doing reduce the stress concentrating effect of the hole) and, subsequent to crack initiation, reduce the stress intensity at the crack tip [15]. In addition, the adhesive could act as an environmental barrier, or provide an interlayer which would reduce the effects of fretting between the rivet and hole surface. The proposal was investigated in two complementary series of fatigue tests which are covered in this chapter.

4.8.1 Specimens and testing program

The basic forms of the two types of fatigue specimens employed in this investigation are shown in Figure 4.5. The use of two types was necessitated by the unavailability of suitable test material. Type (a) was made from offcuts of B.S.L168 aluminium alloy extruded bar 63.5 mm × 31.75 mm in section (Serial GR) which was used for the investigation reported in [21]. Type (b) was taken from offcuts of 32 mm thick 2014-T651 aluminium alloy rolled plate (Serial GJ) used in [20]. In both cases, the longitudinal axis of the specimen was parallel to the direction of extrusion/rolling of the material. The chemical compositions and static properties of these materials are equivalent to those of the French alloy A-U4SG used for the manufacture of the spars.

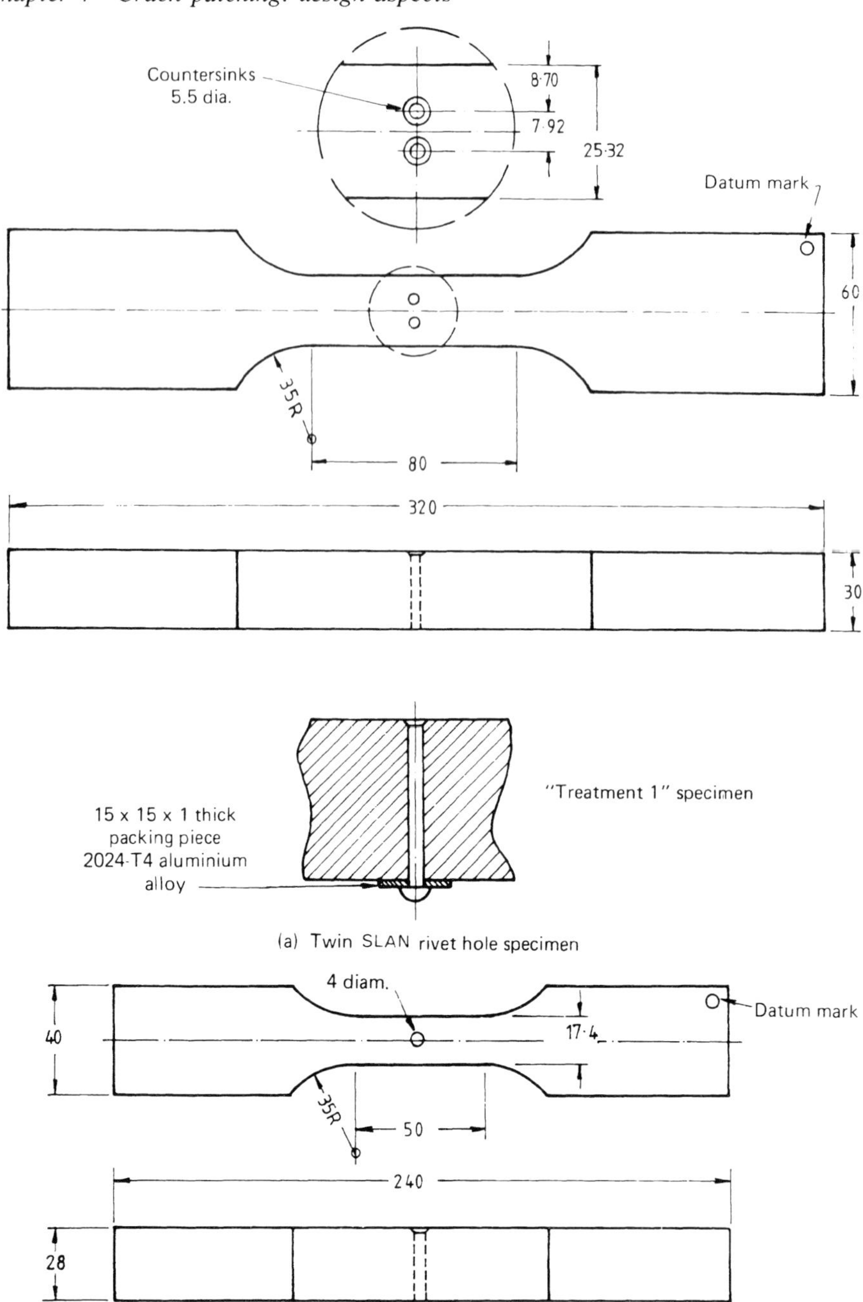

Figure 4.5. Fatigue test specimens (all dimensions in mm).

4.8.2 Type (a) specimens

In these specimens, the pitch between the twin holes was the same as the nominal pitch of the two rivet holes in the SLAN, and the distance from the centre of each hole to the closest side of the specimen was that from the inner side of the 8 mm spar bolt hole to the centre line of the first rivet hole, i.e. 8.70 mm. A specimen thickness of 30 mm was chosen to correspond to the thickness of the spar flange at the SLAN section.

Eight different hole treatments were investigated.

(i) Holes drilled to 3.3 mm diameter and 0.125 inch diameter universal head rivets inserted against a packing piece, with the tail of the rivet peened into the countersink.
(ii) Holes drilled to 3.3 mm diameter and left empty. A condition equivalent to that of rivet removal from the spar flange without any reworking of the rivet holes.
(iii) Holes reamed to 4 mm diameter and left empty. Representing a situation in which the rivet holes were simply cleaned up for inspection.
(iv) Holes as in (iii) incorporating selected 5/32 inch (4 mm) diameter countersink-head (2117 aluminium alloy rivets pressed-in by hand to provide a neat fit. Designed to allow some load transfer through the rivet, but to enable easy rivet removal for hole inspection during testing if required.
(v) Holes and rivets as in (iv), but with rivets permanently bonded in position using CIBA epoxy adhesive Type K138.
(vi) Holes cold-expanded 2.7 to 3.4% using the Boeing split-sleeve process to finish at 4 mm diameter. Holes left empty. The cold-expanding process introduces a residual compressive stress field adjacent to the hole which can retard fatigue crack initiation and growth.
(vii) Holes as in (vi), but incorporating close-fit rivets as in (iv).
(viii) Holes and rivets as in (vii), but with rivets adhesively bonded as in (v).

4.8.3 Type (b) specimens

In these specimens the distance from the centre line of the hole to the side of the specimen (8.70 mm) corresponded to that in the Type (a) specimens. The specimen thickness was, however, slightly less i.e. 28 mm. Four hole treatments were investigated.

(i) Holes reamed to 4 mm diameter and left empty. Equivalent to Type (aiii).
(ii) Holes reamed as in (i) and left empty, but a coating of adhesive applied to the hole surface.
(iii) Holes reamed as in (i), but incorporating pressed-in close-fit rivets. Equivalent to type (aiv).
(iv) Holes reamed as in (i), but incorporating adhesive-bonded close-fit rivets. Equivalent to Type (av).

4.8.4 Fatigue tests

The multi-load-level fatigue testing sequence used for this investigation was identical to that used for the other Mirage life-enhancement programs [18–20, 21]. It consisted of a 100-flight sequence of four different flight types. Cycles of +6.5 g/−1.5 g and +7.5 g/−2.5 g (a total of 39 cycles in 100 flights) were applied at a cyclic frequency of 1 Hz, whereas the remaining 1950 cycles per 100 flights were at 3 Hz. Sine-wave loading was adopted throughout.

For Type (a) specimens, fatigue loads were based on the assumption that +7.5 g corresponded to a gross-area stress of 235 MPa and that there was a linear stress/g relationship, i.e. the 1g gross-area stress was 31.3 MPa. Specimens with 3.3 mm diameter holes had a nominal nett area of 562 mm^2 while those with 4 mm holes had a nett area of 520 mm^2. The resulting nett area stresses were 318 MPa and 344 MPa respectively – a difference of about 8%.

The nett area stress chosen for the Type (b) specimens was the same as that for Type (a) specimens incorporating 4 mm holes, i.e. 344 MPa. In this case, the gross area stress was 265 MPa.

Tables 4.6 and 4.7 give the individual fatigue lives, log. average lives and standard deviations of log. life for the various groups of Type (a) and Type (b) specimens.

4.8.5 Discussion

The individual fatigue test series involving the twin hole and single hole specimens both demonstrate the effectiveness of adhesively-bonded rivets in providing a significant increase in the life to failure relative to those for other hole treatments. Compared with specimens having reamed open holes, the ratio of lives for specimens incorporating adhesively-bonded rivets in reamed holes are 2.65 and 2.61 for the twin-hole and single-hole specimens respectively. On the limited data available, the cold-expansion of the holes in the twin-hole specimens did not result in a significant increase in life compared with that of reamed open-hole specimens, but bonding of rivets in cold-expanded holes again provided a marked increase in life. Furthermore, a coating on the hole surface of the single-hole specimens did not result in an improvement in life. For both types of specimens the lives of those incorporating pressed-in rivets were not significantly different to those of open-hole specimens with similar hole conditions.

Three reasons which could be advanced for the significant improvement in life associated with the use of adhesively-bonded rivets are:

(a) the adhesive acts as a barrier to inhibit crack initiation which might otherwise have been accelerated by environmental interaction;
(b) the adhesive acts as non-metallic interlayer, thus separating the rivets and hole surface and reducing the potentially deleterious effects of fretting;
(c) the adhesive provides improved load transfer characteristics at the section, both before and after crack initiation.

The tests on the single-hole specimens show that the adhesive coating, as such, does not play a major part in the increased life associated with adhesively-bonded rivets see Figure 4.6. An examination of the fracture surfaces of specimens with filled holes (i.e. incorporating either pressed-in or adhesively-bonded rivets) indicated the presence of fretting at, or close to, the countersink-end of nearly every hole, with lesser or no fretting at the other end of the holes. Of the 20 'filled-hole'

Table 4.6. Fatigue test results, type (a) specimens.

Hole treatment	Specimen No. GR	Life (flights)
(i) Drilled 3.3 mm, filled 1/8 inch rivets	18E	9,542
	13B	10,042
	19C	10,320
	log. average life = 9,963	
	s.d. log. life = 0.017	
(ii) Drilled 3.3 mm, open holes	26B	7,500
	16E	8,105
	log. average life = 7.797	
	s.d. log life = 0.024	
(iii) Reamed 4 mm, open holes	25B	5,542
	18B	6,342
	log. average life = 5,929	
	s.d. log life = 0.041	
(iv) Reamed 4 mm, pressed-in 5/32 inch rivets	17B	3,735
	23B	3,842
	14B	6,335
	log. average life = 4,496	
	s.d. log life = 0.129	
(v) Reamed 4 mm, adhesively-bonded 5/32 inch rivets	20B	14,642
	13C	16,040
	16B	16,440
	log. average life = 15,688	
	s.d. log. life = 0.026	
(vi) Expanded 4 mm, open holes	19B	7,142
	24B	9,442
	log. average life = 8,212	
	s.d. log. life = 0.086	
(vii) Expanded 4 mm, pressed-in 5/32 rivets	17C	5,942
	14E	7,642
	21B	8.080
	log. average life = 7,159	
	s.d. log. life = 0.071	
(viii) Expanded 4 mm, adhesively bonded 5/32 inch rivets	15B	10,542
	22B	23,542
	15C	26,130
	log. average life = 18,648	
	s.d. log. life = 0.216	

Table 4.7. Fatigue test results, type (b) specimens.

Hole treatment	Specimen No. GJ	Life (flights)
(i) Reamed 4 mm, open holes	IU	2,542
	IZB	3,242
	log. average life = 2,871	
	s.d. log. life = 0.075	
(ii) Reamed 4 mm, open holes coated with adhesive	IY	2,742
	IV	2,842
	log. average life = 2,792	
	s.d. log. life = 0.011	
(iii) Reamed 4 mm, pressed-in rivet	2A16	3,542
	IZ	3,820
	log. average life = 3,678	
	s.d. log. life = 0.023	
(iv) Reamed 4 mm, adhesively-bonded rivet	IX	5,742
	IZA	7,942
	2B16	8,942
	log. average life = 7,415	
	s.d. log. life = 0.100	

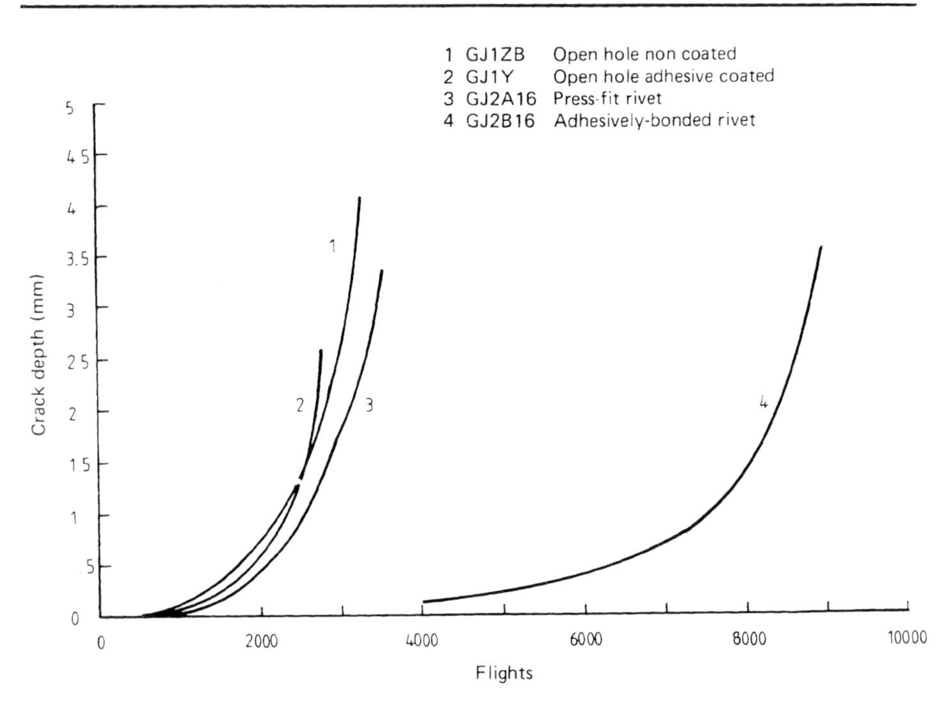

Figure 4.6. Crack propagation, type (b) specimens.

specimens (15 twin-hole and five single-hole), the primary crack initiation in 13 was some distance from the ends of the hole, and in the other four, close to the end opposite to the countersink. It was only in the other three cases (one twin-hole cold-expanded, adhesively-bonded specimen No. GR22B; and two single-hole – one each with pressed-in and adhesively-bonded rivets, GJ1Z and GJ1ZA respectively) that the primary crack developed from close to the countersink and fretting apparently did play a significant part in its initiation. Thus, as fretting was not a major factor in crack initiation in the non-adhesively-bonded specimens, the anti-fretting properties of the adhesive interlayer are not responsible for the benefits resulting from the use of adhesively-bonded rivets.

4.8.6 *Numerical analysis*

In order to understand the mechanisms which resulted in the increase in fatigue life of the adhesively-bonded specimen a detailed finite element analysis was undertaken on the single-hole specimen containing a bonded rivet. Initially, the specimen was considered to be uncracked, and due to symmetry, only a quarter of it was modelled. The resultant plane strain finite element model consisted of 38 eight-noded isoparametric quadrilateral elements and eight six-noded isoparametric triangular elements – see Figure 4.7. Element stiffness matrices were computed using reduced integration and double precision and the solution was also performed using double precision.

The aluminium alloy rivet and specimen were both assumed to have a Modulus of Elasticity of 73 000 MPa and a Poisson's ratio of 0.32, while the adhesive was assumed to have a Modulus of Elasticity of 700 MPa and a Poisson's ratio of 0.35. The adhesive layer was very thin and although its thickness was not precisely known, it was, for the purpose of this analysis, taken to be 0.0127 mm. It was assumed that the applied tensile stress corresponded to the maximum load of 7.5 g in the fatigue sequence. This resulted in a gross area 'applied' stress of 265 MPa, and a nett section stress at the hole of 344 MPa.

Unfortunately, no information on the static tensile or fatigue strength of the adhesive was available and so two separate analyses were undertaken where:

(i) the adhesive was assumed to not yield or fail,
(ii) the adhesive was assumed to fail wherever the peel stress was tensile – this represented a condition in which at least 50% of the glue line had failed.

The results of these analyses can be found in Table 4.8 as can the result for the case when the hole was unfilled. This shows that in each case, the adhesively-bonded rivet has dramatically reduced the stresses around the hole, and thus a significant increase in the life to crack initiation could be expected.

Attention is now directed to the situation when the hole is cracked. Cracks of various discrete lengths, up to a maximum of 2.5 mm, were considered for the following cases:

(i) a through crack on one side of the hole only (half the structure modelled),
(ii) through cracks of equal length, on each side of the hole (a quarter of the structure modelled).

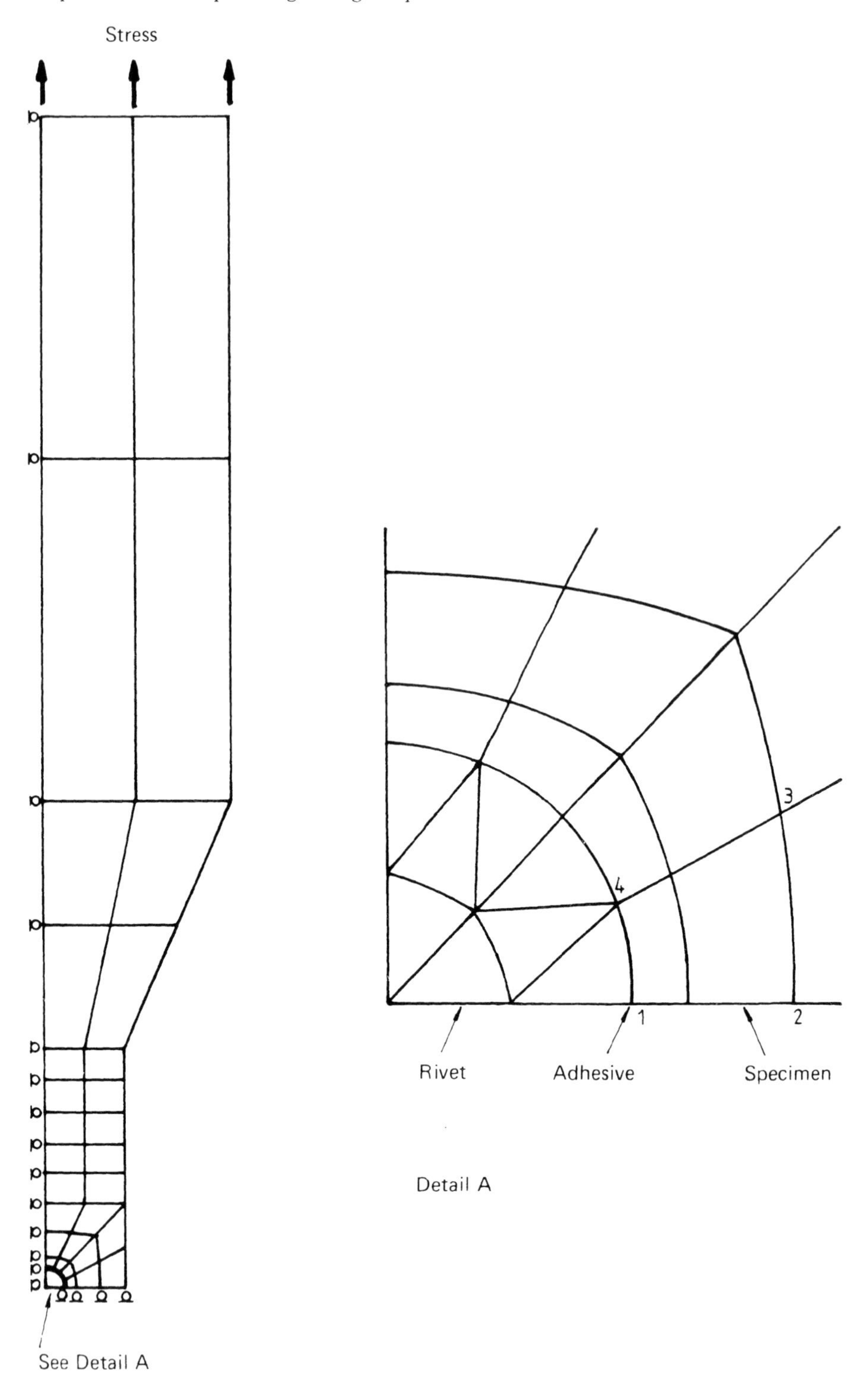

Figure 4.7. (a) Finite element mesh for uncracked specimen – bonded rivet.

Table 4.8. Maximum principal stress for the uncracked specimen.

Case considered	Stress (MPa)
Unfilled hole	853
Bonded rivet (no adhesive failure)	337
Bonded rivet (adhesive failed in tension)	427

The finite element model for these problems consisted of approximately 74 eight-noded isoparametric quadrilateral elements and 32 six-noded isoparametric triangular elements for case (i) and 36 and 24 respectively for case (ii). As before, reduced integration was used and the solution was performed in double precision. The results of this analysis are given in Tables 4.9 and 4.10. Also given are the values of the stress intensity factor for the case when the hole is unfilled, and for the case when there is a crack but no rivet hole.

There are a number of important points indicated by this work:

(i) Adhesive bonding significantly reduces both the local stress concentration at the hole and the stress intensities at the crack tips, thus retarding crack initiation and reducing fatigue crack propagation rates.

(ii) The values of stress intensity factors for the case of a crack on one side and for the case of a crack of the same length on both sides of the hole are almost identical.

(iii) As the crack length increases, a crack at a bonded rivet hole behaves as if the specimen does not contain a hole. In the present case this asymptotic behaviour is effectively reached at a crack length of approximately 2.5 mm. This is particularly important since it allows simple, and yet accurate,

Table 4.9. Stress intensity factors K ($MPa\sqrt{m}$) for a crack hole – crack on one side only.

Crack length	Bonded rivet		No hole	Unfilled hole
	No adhesive failure	Adhesive failed in tension		
0.5	9.1	12.1	–	25.6
0.9	11.3	15.0	–	27.8
1.5	14.3	18.4	13.1	29.9
2.0	15.9	20.1	14.8	30.9
2.3	17.1	21.4	16.02	32.1

Table 4.10. K ($MPa\sqrt{m}$) for a crack on both sides.

Crack length	Bonded rivet		Unfilled hole
	No adhesive failure	Adhesive failed in tension	
0.5	9.1	12.2	26.4
0.9	11.3	15.3	29.7
1.5	14.5	18.8	33.9
2.3	17.6	22.6	39.2
2.5	–	23.6	–

analytical estimates to be obtained for stress intensity factors of cracked holes which are to be repaired by a bonded insert.

Whilst this work has concentrated on bonded rivets with a particular adhesive thickness, a more detailed investigation into the effects of variable adhesive thickness is currently underway.

4.9 Repair of cracked fastener holes

In the previous section, the fastener hole was not initially cracked. To investigate the behaviour of cracked fastener holes, a limited series of tests were performed on 2024-T4 aluminium alloy plates. The specimens were 72 mm wide, 11.2 mm thick and had a working length of 200 mm. The specimens contained a centrally located hole 10 mm in diameter which in turn contained two diametrically opposed 3 mm radius quadrant cracks on one face of the plate (see Figure 4.8).

Three of these specimens were tested at a constant amplitude cyclic tensile stress of 68.9 MPa with stress ratio $R = 0.01$. The fatigue lives to failure for these specimens are given in Table 4.11.

Four additional specimens were repaired with a 1 mm thick bonded steel sleeve using an acrylic adhesive (viz: Flexion 214) and tested under the same load spectrum. Their fatigue lives are also given in Table 4.11.

Although there is scatter in the comparative fatigue lives the results show a significant increase of 114% in the average fatigue life of the repaired holes.

In the case of the repaired holes, the scatter is believed to be primarily due to the bonding process. Further development of this process is currently underway and can be expected to yield more consistent results.

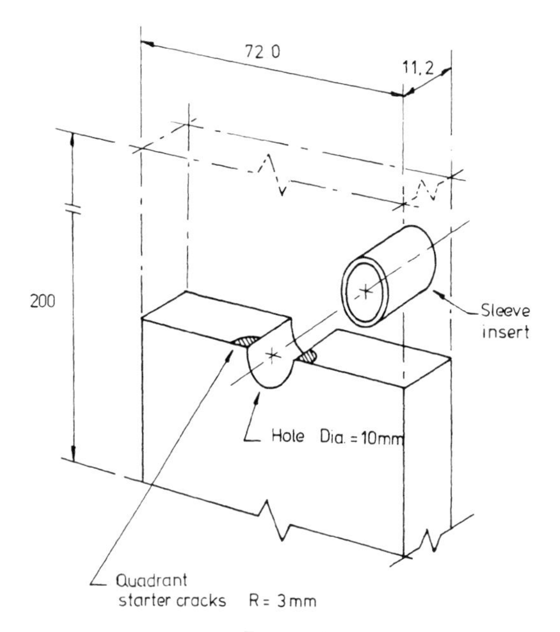

Figure 4.8.

Table 4.11. Experimental fatigue test results for specimen with two quadrant cracks at hole.

Specimen condition	Specimen number	Fatigue life (cycles)
Unrepaired	BP74AK	97,600
	BP74AE	86,600
	BP74AA	100,700
Repaired	BP74AB	162,500
	BP74AC	147,000
	BP74AJ	258,000
	BP74AD	248,100

References

[1] Dowrick, G., Cartwright, D.J. and Rooke, D.P., The effects of repair patches on the stress distribution in a cracked sheet, Royal Aircraft Establishment Technical Report 80098, (August 1980).

[2] Dowrick, G. and Cartwright, D.J., The effect of a circularly reinforcing patch on a crack in a uniaxially stressed sheet, Report ME/80/2, Southampton University, (January 1980).

[3] Ratwani, M.M., Analysis of cracked adhesively bonded structures, AIAA/ASME 19th Structures, Structural Dynamics and Materials Conference, pp. 155–163, (1978).

[4] Ratwani, M.M., Development of bonded composite patch repairs for cracked metal structure, Proc. International Workshop on defence applications of advanced repair technology for metal and composite structures, Naval Research Laboratories, Washington 22–24th July 1981, pp. 370–403, (1981).

[5] Jones, R. and Callinan, R.J., Finite element analysis of patched cracks, *J. Structural Mechanics*, 7, 2, pp. 107–130, (1979).

[6] Jones, R. and Callinan, R.J., A design study in crack patching, *J. Fibre Science and Technology*, 14, pp. 99–111, (1981).

[7] Rose, L.R.F., A cracked plate repaired by bonded reinforcements, *Int. J. Fracture*, 18, 135–144 (1982).

[8] Baker, A.A., Roberts, J.D. and Rose, L.R.F., Use of overlap joint parameters in estimating the K reduction due to crack patching, Proc. Int. Workshop on defence applications of advanced repair technology for metal and composite structures, Naval Research Labs., Washington 22–24th July 1981, pp. 335–354, (1981).

[9] Jones, R., Davis, M., Callinan, R.J. and Mallinson, G.D., Crack patching: analysis and design, *J. Structural Mechanics* 10(2), pp. 177–190, (1980).

[10] Baker, A.A., Callinan, R.J., Davis, M.J., Jones, R. and Williams, J.G., Application of BFRP crack patching to Mirage III aircraft, Proc. 3rd Int. Conf. on Composite Materials, Vol. II, Paris (1980), pp. 1424–1438.

[11] Zienkiewicz, "The finite element method in engineering science", McGraw-Hill, London (1971).

[12] Jones, R. and Callinan, R.J., Thermal considerations in the patching of metal sheets with composite overlays, *J. Structural Mechanics*, 8(2), pp. 143–149, (1980).

[13] Jones, R., Neutral axis offset effects due to crack patching, *J. Composite Structures*, 1, 2, pp. 163–174, (1983).

[14] Jones, R., Callinan, R.J. and Aggarwal, K.C., Analysis of bonded repairs to damaged fibre composite structures, *Engng. Frac. Mechanics*, 17, 1, 37–46 (1983).

[15] Jones, R. and Callinan, R.J., New thoughts on stopping cracks which emanate from holes, *Int. J. Fracture*, 17, R53–R55, (1981).

[16] Callinan, R.J., Residual Strength of a cracked lug, A.R.L. Structures Tech. Note 422, Aeronautical Research Laboratories, Australia, (1977).

[17] Mann, J.Y., Pell, R.A., Jones, R. and Heller, M., The use of adhesive bonded rivets to reduce the effects of rivet holes on fatigue life, ARL Structures Report 399, (1984).

[18] J.Y. Mann, R. Kalin and F.E. Wilson, Extending the fatigue life of a fighter aircraft wing. Aircraft Fatigue in the Eighties. (Editors: J.B. deJonge and H.H. van den Linden). Netherlands, National Aerospace Laboratories, 1981, pp. 1.7/1–1.7/42.
[19] J.Y. Mann, A.S., Machin and W.F. Lupson, Aero, Res. Labs, Structures Report 398, January 1984.
[20] J.Y. Mann, G.W. Revill and W.F. Lupson, Improving the fatigue performance of thick aluminium alloy bolted joints by cold expansion and the use of interference fit steel bushes, Aero. Res. Labs, Structures Note 486, April 1983.
[21] J.Y. Mann, A.S. Machin, W.F. Lupson and R.A. Pell, The use of interference fit bolts or bushes and cold expansion for increasing the fatigue life of thick section aluminium alloy bolted joints, Aero. Res. Labs, Structures Note 490, August 1983.

Appendix A

As shown in section 4.2, equation 4.3, it is necessary for the displacement field to be at least a quadratic in z. Hence when performing a three-dimensional analysis the twenty-noded isoparametric brick or the fifteen noded isoparametric wedge elements must be used when modelling the patch, adhesive and the sheet. In order to avoid numerical ill-conditioning, which may arise due to the large aspect ratios of the elements used to model the adhesive layer, it is necessary to evaluate the stiffness matrix using either a $2 \times 2 \times 2$ or a $2 \times 2 \times 3$ with 3 Gauss points in the z direction, array of Gauss points instead of the $3 \times 3 \times 3$ array normally used.

On computers which use less than thirty-six bit words the formulation of the element stiffness matrices and the solution should be performed in double precision.

If the problem does not involve bending due to neutral axis offset effects then it is possible to use twenty-noded isoparametric elements for the adhesive and membrane elements for the patch and the skin. To make allowance for the shear deformation in the two adherends it is necessary to use for the adhesive G_a^{eff}, where

$$G_a^{\text{eff}} = 1/[1/G_a + 3tp/8t_aG_a + 3tr/8t_a\mathrm{G}_R]$$

This approach is identical to that outlined in section 4.2 and has the advantage of using only standard elements.

L.R.F. ROSE

5

Theoretical analysis of crack patching

5.1 Introduction

The development of high-strength fibres and adhesives has made it possible to repair cracked metallic plates (for example, cracked portions of an aircraft's fuselage) by bonding reinforcing patches to the plate over the crack. The process is referred to as crack patching, and several repairs using this procedure are documented in Chapter 6, Section 7, where the practical advantages over the traditional methods of riveting or welding metallic reinforcements are discussed. The aim of this chapter is to discuss some basic theoretical aspects of the mechanics of bonded reinforcements. To that end, we shall consider a relatively simple repair configuration, namely a centre-cracked plate which is repaired by an elliptical patch and which is subjected to a remote biaxial stress, as illustrated in Figure 5.1. It will be shown that by using simplifying assumptions which are based on an analysis of the load-transfer characteristics of bonded joints, one can derive analytical estimates for the following quantities which are of primary interest in assessing the efficiency and the viability of the repair:

(i) the reduction in the stress intensity factor,
(ii) the maximum shear strain in the adhesive,
(iii) the maximum tensile stress in the reinforcing patch,
(iv) the change in overall stiffness due to the crack and the bonded reinforcement.

The fundamental idea is to divide the analysis into two stages [1]. First we consider the redistribution of stress which would be caused by the reinforcement if it were bonded to an *uncracked plate*. The quantity of interest is the normal stress σ_0 in the plate. At the second stage, we make a cut in the plate and allow the stress σ_0 to relax to zero. The main problem then is to determine the resulting stress intensity factor,

The value of dividing the analysis into these two stages is that different simplifying assumptions are appropriate for each stage. Thus, for stage I it can be assumed that the adhesive bond does not allow any relative displacement between the plate and the reinforcement. This rigid-bond assumption is appropriate when the actual

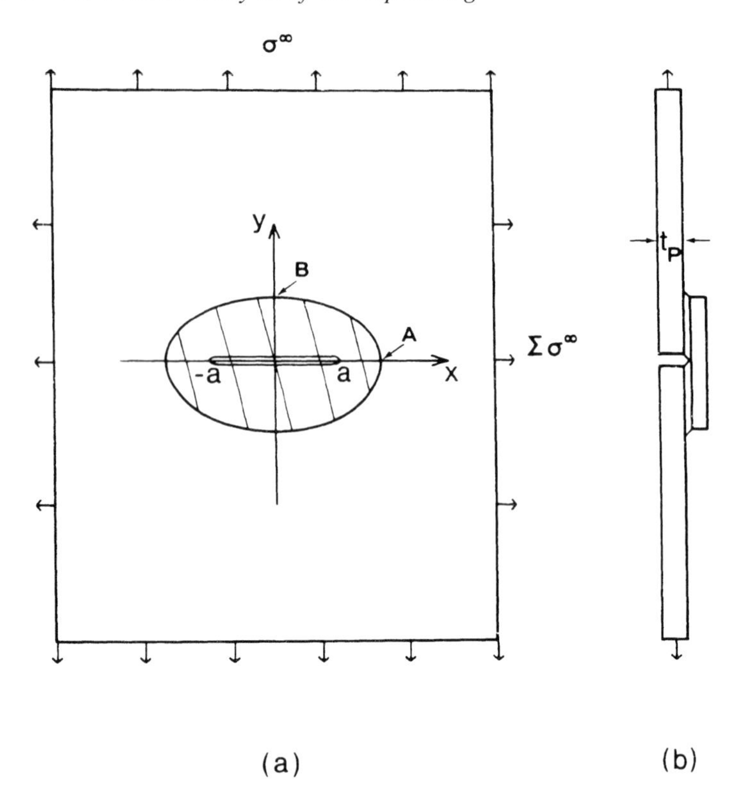

Figure 5.1. Repair configuration: (a) Plan view, (b) Cross-section along centre line ($x = 0$).

width of the load-transfer zone around the boundary of the reinforced region is small compared with the in-plane dimensions of the reinforcement. The width of this load-transfer zone can be estimated from the theory of bonded joints (Section 5.3). For the second stage of the analysis, the assumption of a rigid bond must be discarded, as it would imply that the crack cannot open. It is now important to model more accurately the transfer of load through the adhesive layer in the immediate vicinity of the crack, but both the plate and the reinforcement can now be assumed to be of infinite extent. The crucial feature to emerge is that for a sandwich structure consisting of a centre-cracked plate bonded to an uncracked reinforcing plate, the stress intensity factor K does not increase indefinitely with increasing crack length, as it would if there were no reinforcement. Rather, K approaches asymptotically a limiting value which can be estimated by a very simple calculation [2].

The plan of this Chapter is as follows. The basic problem which will be considered is formulated in Section 5.2, while Section 5.3 summarizes some results of the theory of bonded joints which will be required in the sequel. The two-stage analysis outlined above is presented in more detail in Section 5.4. The following sections deal with two problems which can be important in practice, namely the residual thermal stress due to the adhesive curing process (Section 5.5) and the effect of out-of-plane bending due to one-sided reinforcement (Section 5.6). In

Section 5.7, a recent extension of the basic theory [1, 2] to the case of partial reinforcement is described, in which the reinforcing action for stage II is represented by distributed springs acting across the crack faces. Numerical results are presented in non-dimensional form for the reduction in K which can be achieved by partial reinforcement, and these results are compared with experimental measurements (described in Chapter 6) for one particular case.

5.2 Formulation and notation

Consider an infinite centre-cracked plate, with the crack along the line segment $|x| \leqq a$, $y = 0$, as in Figure 5.1. The plate is repaired by a patch bonded to one face of the plate over an elliptical region $\mathscr{D}$, defined by

$$\mathscr{D} = \{(x, y);\, (x/A)^2 + (y/B)^2 \leqq 1\}, \tag{5.1}$$

which completely covers the crack ($A > a$). After this repair, the plate is subjected to a remote biaxial stress specified by

$$\begin{aligned} \sigma^P_{yy}(x^2 + y^2 \to \infty) &= \sigma^\infty_{yy} \equiv \sigma^\infty, \\ \sigma^P_{xx}(x^2 + y^2 \to \infty) &= \sigma^\infty_{xx} \equiv \Sigma\sigma^\infty, \ (\Sigma < 1), \\ \sigma^P_{xy}(x^2 + y^2 \to \infty) &= 0. \end{aligned} \tag{5.2}$$

The problem to be considered is that of calculating

(i) the *stress intensity factor* K_r in the repaired plate,

$$K_r = \lim_{x \to a+} \{2\pi(x - a)\}^{1/2} \sigma^P_{yy}(x, y = 0), \tag{5.3}$$

(ii) the *maximum tensile stress* in the reinforcing patch, $\sigma^R_{\max}$,
(iii) the *maximum shear strain* in the adhesive $\gamma^A_{\max}$.

Notation.
Subscripts or superscripts P, R, A will be used to identify parameters pertaining to the plate, the reinforcing patch or the adhesive layer, respectively. Thus E_P, E_R will denote the Young's moduli of the plate and the reinforcement, G_A the shear modulus of the adhesive, and t_P, t_R, t_A the respective thicknesses.

Assumptions.
To bring out more clearly the essential features of the present approach, it will be convenient to begin with the following simplifying assumptions, which can later be relaxed once the simplified problem has been solved.

(i) The plate and the reinforcement are both isotropic and have the same Poisson ratio $\nu (= \nu_P = \nu_R)$.
(ii) There is no residual thermal stress induced by bonding.
(iii) The reinforced plate is smoothly constrained against the out-of-plane bending (due to the one-sided reinforcement).

(iv) All deformations are linearly elastic; in particular, the adhesive remains elastic, and any plastic deformation around the crack tip is restricted to a small zone (compared with the crack length) so that it can be ignored when calculating K_r.

(v) The elastic state of the plate and the reinforcement can be idealized as states of generalized plane stress, ignoring therefore any variation across the thickness.

(vi) The adhesive layer acts as a shear spring, as in the classical theory of bonded joints (see Section 5.3).

Basic Parameters.

Because of assumptions (v) and (vi) the elastic moduli and the thicknesses do not occur separately in the analysis. Instead, they appear in the combinations $E_P t_P$, $E_R t_R$, which represent the *stiffness* of the plate and the reinforcement respectively, and G_A/t_A which represents the spring constant of the adhesive layer. One important non-dimensional parameter will therefore be the

$$\textit{stiffness ratio } S = (E_R t_R)/(E_P t_P). \tag{5.4}$$

The spring constant of the adhesive layer enters the analysis via the load-transfer length β^{-1} which will be discussed in the next section.

In the absence of a reinforcing patch, the normal stress σ_{xx} parallel to the crack would not affect the stress intensity factor, but this is no longer true for the repaired plate. After the repair, K_r will depend on the *applied-stress ratio* Σ defined by equation (5.2), and also on the *aspect ratio* B/A of the reinforcement. Because of the assumption of elastic deformation (assumption (iv) above), the main unknowns K_r, σ^R_{max}, γ^A_{max} will all depend linearly on the *principal applied stress* σ^∞ of equation (5.2). The analytical results which will be derived in Section 5.4 will show clearly how these unknowns depend on the basic non-dimensional parameters of the repair configuration (namely S, B/A and Σ) and on the load-transfer length β^{-1}.

Typical values.

To assess whether the simplifying assumptions which are required in the present analysis are likely to be justified in practice, we shall use the following as typical values for the physical parameters.

$$E_P = 70\,\text{GPa},\ t_P = 3\,\text{mm},$$

$$E_R = 200\,\text{GPa},\ t_R = 1\,\text{mm},$$

$$\nu_P = \nu_R = \nu = 1/3,$$

$$G_A = 0.7\,\text{GPa},\ t_A = 0.2\,\text{mm}.$$

The *crack length* a will be taken to be 25 mm, as in the experimental (edge-cracked) specimen described in Chapter 6, Section 6, which is representative of actual repairs.

5.3 Load transfer to bonded reinforcements

Consider the simple reinforcement configuration shown in cross-section in Figure 5.2(a), in which a reinforcing strip of length $2B$ and stiffness $E_R t_R$ is bonded to a strip of length $2L(L > B)$ and stiffness $E_P t_P$, both strips being of *unit width* along the x-axis. The choice of axes is designed to correspond with that used for the repair configuration in Figure 5.1. The stresses and displacements in this reinforced strip can be calculated explicitly using the conventional one-dimensional (1D) theory of bonded joints (see, for example, [3]), which is based on the following assumptions.

(i) Each adherend is treated as a one-dimensional elastic continuum whose deformation is specified by longitudinal displacement u and a longitudinal tensile stress σ. The stress-displacement relations for the plate and reinforcement respectively are

$$\sigma_P(y) = E_P u_P'(y), \sigma_R(y) = E_R u_R'(y) \tag{5.4}$$

where the dash denotes a differentiation with respect to y.

(ii) The adhesive layer acts as a shear spring with the adhesive shear stress τ_A given by

$$\tau_A(y) = (G_A/t_A)\{u_P(y) - u_R(y)\}. \tag{5.6}$$

(iii) The shear tractions exerted by the adhesive can be replaced by an equivalent body force distributed uniformly across the thickness of each adherend, leading to the differential equilibrium equations

$$t_P\sigma_P'(y) = -t_R\sigma_R'(y) = \tau_A(y). \tag{5.7}$$

These assumptions lead to the differential equation

$$\tau_A''(y) - \beta^2\tau_A(y) = 0, \tag{5.8}$$

$$\beta^2 = \frac{G_A}{t_A}\left\{\frac{1}{E_P t_P} + \frac{1}{E_R t_R}\right\} \tag{5.9}$$

For the configuration of Figure 5.2(a) we require τ_A to be odd in y, and $\sigma_R(y = \pm B) = 0$, which implies, by virtue of equations (5.5) and (5.6), that

$$\tau_A'(y = \pm B) = (G_A/t_A)\{F/(E_P t_P)\},$$

where F denotes the known force (per unit width along the x-axis) applied to the ends $y = \pm L$. The appropriate solution of equation (5.8) is therefore

$$\tau_A(y) = \frac{F}{E_P t_P}\frac{G_A}{t_A}\frac{\sinh \beta y}{\beta \cosh \beta B} \tag{5.10}$$

The important feature of this result is that for $\beta B \ll 1$, the adhesive shear stress decays exponentially from the ends ($y = \pm B$) of the overlap, as sketched in Figure 5.2(b); i.e. the load transfer effectively occurs over a length of order β^{-1} at the ends of the overlap.

By integrating equation (5.10) twice to derive the displacement $u_P(y)$, one can

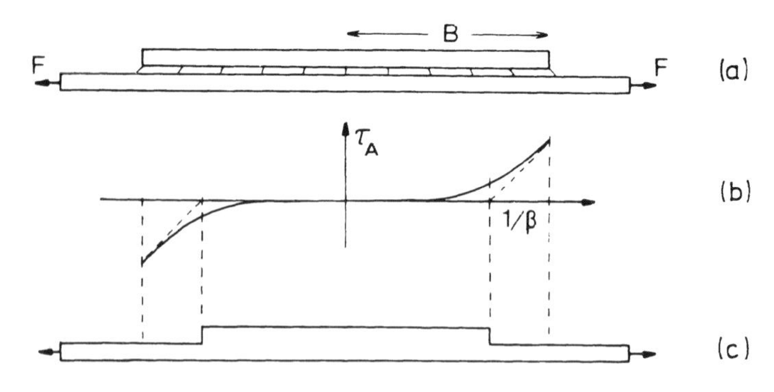

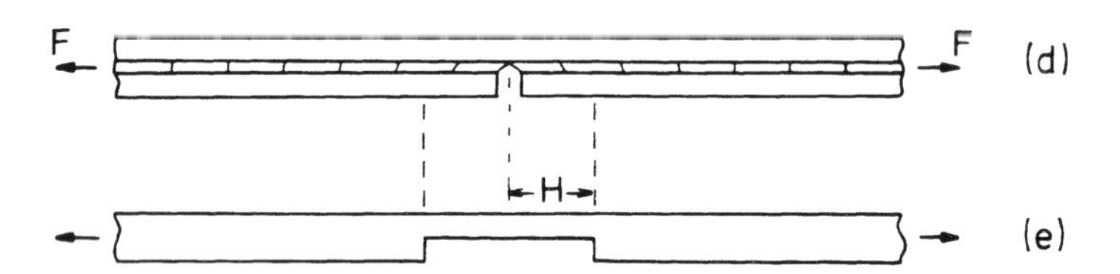

Figure 5.2. Load transfer length for bonded reinforcements: (a) Cross-section for bonded strips; (b) Adhesive shear stress distribution; (c) Equivalent rigid-bond configuration; (d) Cross-section for overlap joint; (e) Rigid-bond configuration equivalent to (d).

readily verify that the overall stiffness between the end points $y = \pm L$ of the reinforced strip in Figure 5.2(a) is the same as that of a strip with a step-change in stiffness from $E_P t_P$ to $E_P t_P + E_R t_R$ over a central portion $|y| \leqq B - b$, as indicated in Figure 5.2(c), with b given by

$$\begin{aligned} \beta b &= \tanh(\beta B) \\ &\cong 1, \quad \text{for} \quad \beta B \gg 1. \end{aligned} \tag{5.11}$$

This equivalence will be exploited in Section 5.4 to assess the redistribution of stress due to a bonded reinforcement.

Energy Variation.
An important calculation which will be required in the sequel is that of determining the change in configurational energy (i.e. elastic strain energy plus potential energy of the loading mechanism) on going from the configuration shown at Figure 5.2(a) to that shown at Figure 5.2(d). This energy change is numerically equal to the work which can be extracted by making a cut at $y = 0$ in Figure 5.2(a) and allowing the stress

$$\sigma_P(y = 0) \equiv \sigma_0, \tag{5.12}$$

on both faces ($y \to 0+$ and $y \to 0-$) to relax to zero. For a linearly elastic system this energy change (per unit width along the x-axis) is thus given by

$$\delta\mathscr{E} \;=\; -\sigma_0 t_P u_P(y \to 0+). \tag{5.13}$$

Using the 1D theory of joints, one can show that

$$\begin{aligned} u_P(y \to 0+) \;&=\; t_P(\sigma_0 \beta t_A/G_A)\coth(\beta B), \\ &\cong\; t_P \sigma_0 \beta t_A/G_A, \quad \text{for} \quad \beta B \gg 1. \end{aligned} \tag{5.14}$$

The energy change given by equations (5.13) and (5.14) is the same as that involved in going from the configuration of Figure 5.2(a) to that of Figure 5.2(e), which involves a step change in stiffness from $E_R t_R$ for $|y| \leqq H$ to $E_P t_P + E_R t_R$ for $|y| > H$, provided that we choose H to be given by

$$H \;=\; \beta^{-1}. \tag{5.15}$$

This equivalence will be used in Section 5.6 when assessing the influence of out-of-plane bending.

5.4 Two-stage analytical solution

We return now to the solution of the problem formulated in Section 5.2, dividing the analysis into two stages as indicated in Section 5.1.

Stage I: Inclusion analogy.
Consider first the redistribution of stress in an *uncracked plate* due to the local stiffening produced by the bonded reinforcement. The 1D theory of bonded joints (Section 5.3) provides an estimate of the load-transfer length for load transfer from the plate to the reinforcement. If that transfer length β^{-1} is much less than the in-plane dimensions A, B of the reinforcement, we may view the reinforced region as an inclusion of higher stiffness than the surrounding plate, and proceed in the following three steps.

(i) Determine the elastic constants of the equivalent inclusion in terms of those of the plate and the reinforcing patch.
(ii) Determine the stress in the equivalent inclusion.
(iii) Determine how the load which is transmitted through the inclusion is shared between the plate and reinforcement.

Step (ii) can be carried out analytically only for certain simple shapes, which is why we have chosen the reinforcing patch to be elliptical. For that shape, the stress is uniform within the equivalent inclusion, as indicated schematically in Figure 5.3(a), and that uniform stress state can be determined analytically. The results are derived in [1] for the case where both the plate and the reinforcing patch are taken to be orthotropic, with their principal axes parallel to the x–y axes. We shall not repeat here the intermediate details of the analysis but simply recall the results for the particular case where both the plate and the reinforcement are isotropic and have the same Poisson ratio, as assumed in Section 5.2. The reinforced region then has a stiffness $E_P t_P + E_R t_R$. (This is no longer strictly correct if the Poisson ratios

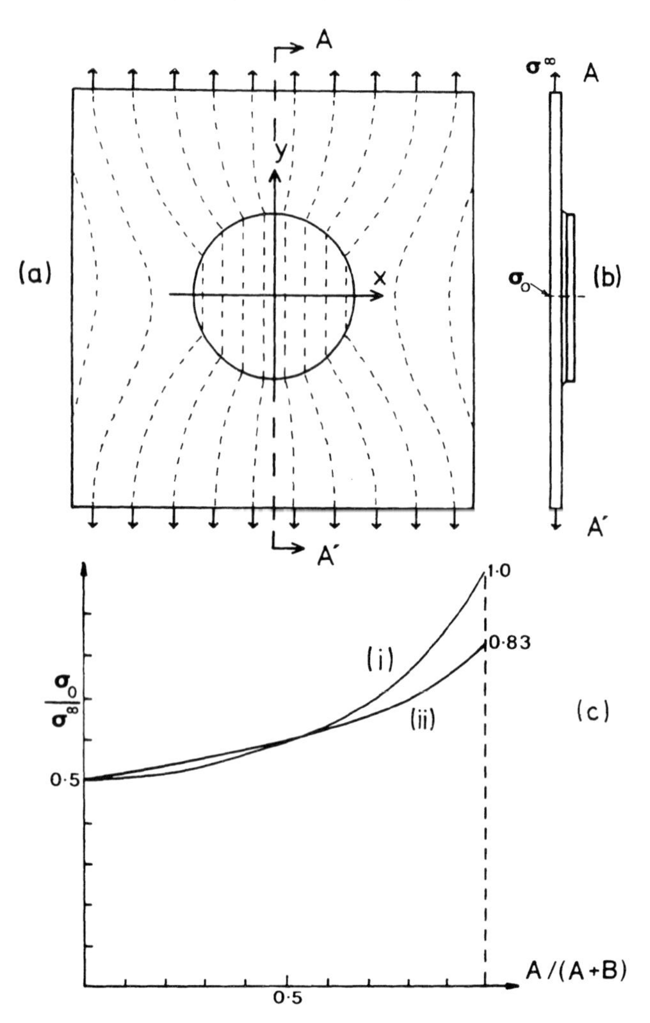

Figure 5.3. Inclusion analogy for Stage I analysis. (a) Flow of load lines into reinforced portion; (b) Cross-section along centre line; (c) Variation of reduced stress with aspect ratio for an elliptical patch of semi-axes A, B, under (i) uniaxial tension ($\Sigma = 0$) and (ii) biaxial tension equivalent to pure shear ($\Sigma = -1$), with $S = 1$, $\nu = 1/3$. For a circular patch, the reduced stress does not depend on the load ratio Σ.

differ). The load transmitted across $y = 0$ within the reinforced region ($|x| < A$) amounts to a force F per unit length along the x-axis, given by

$$F = \sigma^{\infty} t_P \left\{ 1 + \frac{S}{D} [1 + 2(1 + S)B/A(1 - \nu\Sigma) + (1 + S - \nu S)(\Sigma - \nu)] \right\} \tag{5.16}$$

$$D = 3(1 + S)^2 + 2(1 + S)(B/A + A/B + \nu S) + 1 - \nu^2 S^2.$$

This load is transmitted partly through the plate and partly through the reinforcement. The normal stress in the plate will be denoted by σ_0 and is given by

$$\sigma \equiv \sigma_{yy}^P(|x| < A, y = 0) = F/\{t_P(1 + S)\}. \tag{5.17}$$

Thus σ_0 is proportional to the principal applied stress σ^∞, and it depends in addition on the following three parameters: (i) the stiffness ratio S, (ii) the aspect ratio A/B, (iii) the applied-stress ratio Σ. The parameters charactering the adhesive layer do not affect σ_0, but we recall that the idealization used to derive equation (5.16) relies on $\beta^{-1} \ll A$, B, and β^{-1} is of course dependent on the adhesive parameters.

To illustrate the important features of equation (5.17), we show in Figure 5.3(b) the variation of σ_0 with aspect ratio for two loading configurations: (i) uniaxial tension ($\Sigma = 0$), and (ii) equal biaxial tension corresponding to pure shear ($\Sigma = -1$), setting $S = 1$ and $\nu = 1/3$ for both cases. It can be seen that there is little variation for aspects ratio ranging from $B/A = 0$ (horizontal strip) to $B/A = 1$ (circular patch), so that for preliminary design calculations, one can conveniently assume the patch to be circular, to reduce the number of independent parameters. It is also noted that the curves for $\Sigma = 0$ and $\Sigma = -1$ cross over for $B/A = 1$, indicating that, for a circular patch, the transverse stress σ_{xx}^∞ does not contribute to σ_0, so that this parameter can also be ignored in preliminary design estimates.

The inclusion analogy also gives, as a natural by-product, the stress in the plate outside the reinforced region $\mathscr{D}$. The stress at the point $x = 0$, $y = B+$ is of special interest, as it has been observed experimentally that cracking can initiate at that point. This stress is given by

$$\sigma_{yy}^P(0, B+) = F/t_P = (1 + S)\sigma_0. \tag{5.18}$$

Stage II: Upperbounds for K_r.

Once the stress at the prospective crack location is known, one can proceed to the second stage of the analysis in which the plate is cut along the line segment ($|x| < a, y = 0$), and a pressure equal to σ_0 is applied internally to the faces of this cut to make these faces stress-free. Provided that the load transfer to the reinforcement during this second stage takes place in the immediate neighbourhood of the crack, the reinforcement may be assumed to be of infinite extent. Thus the problem at this stage is to determine the stress intensity factor K_r for the configuration shown in Figure 5.4. Without the reinforcement, the stress intensity factor would have the value K_0 given by the well-known formula

$$K_0 = \sigma_0(\pi a)^{1/2} \tag{5.19}$$

This provides an upper bound for K_r, since we must clearly have $K_r < K_0$. However, K_0 increases indefinitely as the crack length increases, whereas the crucical property of the reinforced plate of Figure 5.4(a) is that K_r does not increase beyond a limiting value, denoted by K_∞, as shown in Figure 5.4(c). That limiting value is the value of the stress intensity factor for a semi-infinite crack. It can be determined by deriving first the corresponding crack extension force $\mathscr{G}_\infty$ as follows.

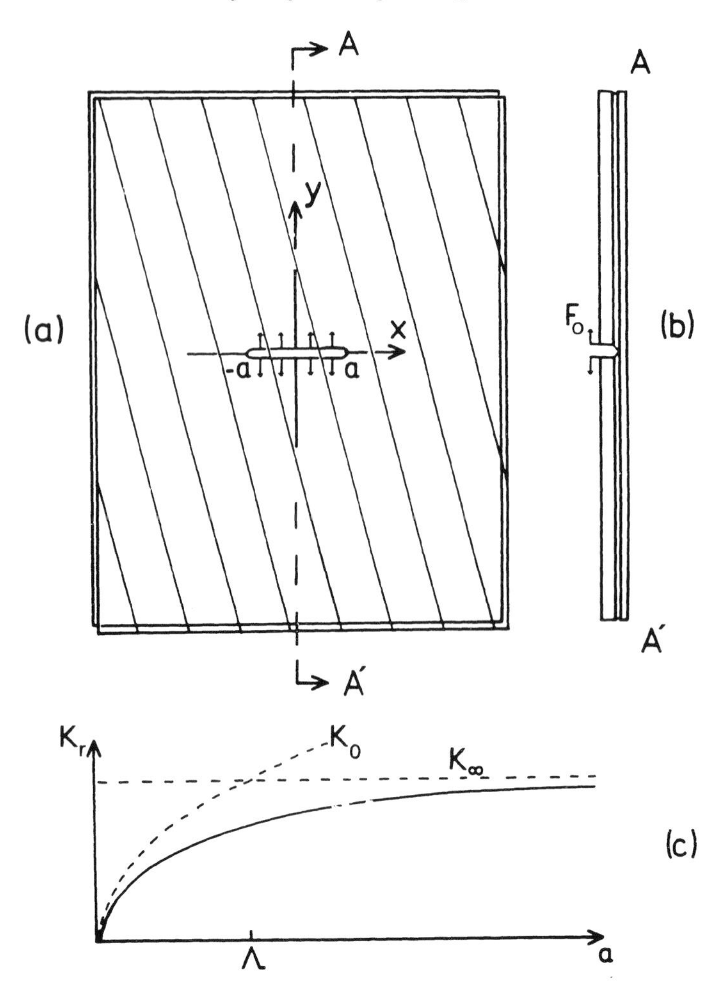

Figure 5.4. Stage II analysis: (a) Plan view, and (b) Cross-section along centre line, of a centre-cracked plate bonded to a reinforcing patch of infinite extent, with tractions applied directly to the crack faces, to cancel the reduced stress calculated during stage I analysis. (c) Variation of K_r with crack length with the two asymptotes shown as dashed lines.

Consider the configuration shown in Figure 5.5. If the semi-infinite crack extends by a distance δa, the stress and displacement fields are simply shifted to the right by δa. The change in energy $\delta\mathscr{E}$ (potential energy of loading mechanism plus strain energy) is the same as that involved in converting a strip of width δa from the state shown as section AA′ in Figure 5.5 to that shown as section BB′. The crack extension force $\mathscr{G}_\infty$ is defined by

$$t_p \mathscr{G}_\infty = \lim_{\delta a \to 0} - (\delta\mathscr{E}/\delta a).$$

To determine $\mathscr{G}_\infty$ therefore, we only need to calculate $\delta\mathscr{E}$, but that is precisely the calculation carried out in Section 5.3, the result being given by equations (5.13) and (5.14). Thus we obtain the following explicit formula for $\mathscr{G}_\infty$:

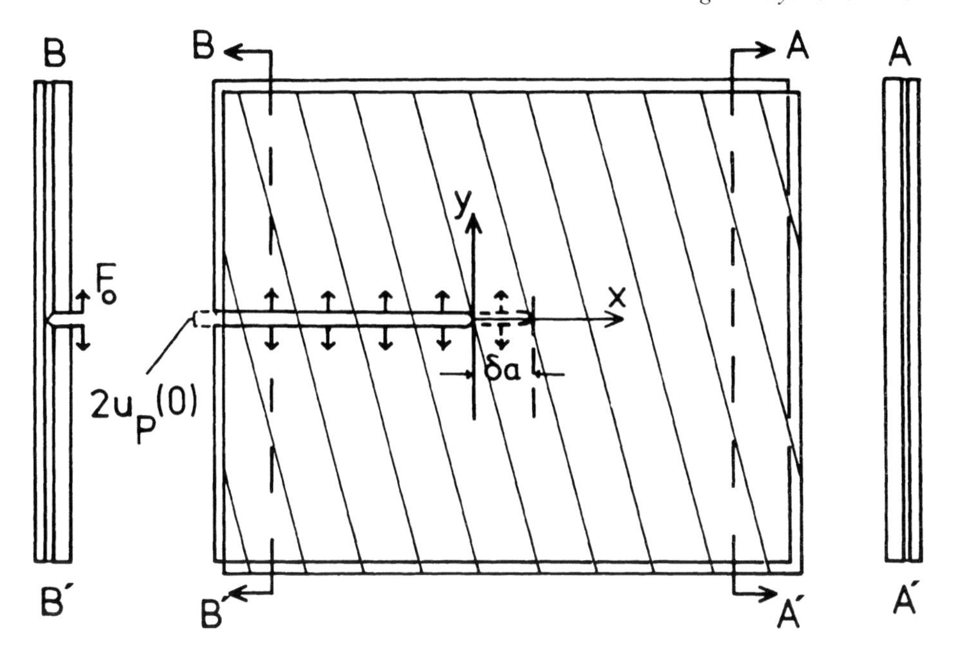

Figure 5.5. Configuration used for calculating the long-crack asymptote for K_r.

$$\mathscr{G}_\infty = (\sigma_0^2/E_P)\pi\Lambda \tag{5.20}$$

$$\pi\Lambda = E_P t_P (t_A/G_A)\beta = (1 + 1/S)\beta^{-1}. \tag{5.21}$$

From equation (5.20), assuming that the usual relation holds between $\mathscr{G}$ and K, (see [2] for a proof), we obtain

$$K_\infty = (E_P \mathscr{G}_\infty)^{1/2} = \sigma_0(\pi\Lambda)^{1/2}. \tag{5.22}$$

It is clear from this derivation that K_∞, like K_0, will also be an upperbound for K_r. It can further be shown that K_0 and K_∞ are in fact the first terms in the asymptotic expansions of K_r in the limits $a/\Lambda \ll 1$ and $a/\Lambda \gg 1$, respectively [2, 4, 5]. By interpolating between these asymptotes one can construct a convenient analytical approximation for K_r,

$$K_r = \sigma_0\{\pi a\Lambda/(a + \Lambda)\}^{1/2}, \tag{5.23a}$$

and hence for the corresponding crack extension force

$$\mathscr{G}_r = K_r^2/E_P. \tag{5.23b}$$

The important feature which emerges from the preceding analysis is the existence of a *characteristic crack length* Λ, which can be derived from the physical parameters of the repair without reference to the actual crack length a or the applied stress σ^∞. It has been noted in [2] that the results of an extensive parametric study [6, 7] all fall on the one curve when plotted against the non-dimensional crack length a/Λ. In the range $a/\Lambda < 1$, the reduction in stress intensity factor is mainly due to the stress reduction from σ^∞ to σ_0 which was calculated at stage I. In the range $a/\Lambda > 1$, another factor becomes increasingly important, namely the restraint on the relative displacement of the crack faces for a given value of σ_0, this relative displacement (or crack opening) cannot exceed the relative displacement which would occur in an overlap joint having the same cross-section. This is the physical basis for the existence of an upperbound K_∞ to K_r and hence the existence of a characteristic length Λ. Using the typcial parameter values of Section 5.2, we find

$$\beta^{-1} = 5.4\,\text{mm}, \quad \Lambda = 3.5\,\text{mm}.$$

It is clear that for the typical crack length $a = 25\,\text{mm}$, K_r should be close to the limiting value K_∞, since $a/\Lambda = 7$.

To estimate the fractional reduction in K which can be achieved for this case we need to have an estimate for σ_0/σ^∞ from stage I of the analysis. Then

$$K_r/K_{\text{initial}} = (\sigma_0/\sigma^\infty)(\Lambda/a)^{1/2}. \tag{5.24}$$

For example with a circular patch ($A/B = 1$) of the same stiffness as the plate ($S = 1$), and under uniaxial tension ($\Sigma = 0$) we have $\sigma_0/\sigma^\infty = 0.6$, for $\nu = 1/3$, so that the fractional reduction in K which can be achieved for the above values of a and Λ is 0.22, i.e. a reduction by a factor of 4.5.

The fundamental simplification which has been achieved is that the configurations AA′ and BB′ in Figure 5.5 represent simple overlap joints, which are much easier to study, both theoretically and experimentally, than the original configuration of a cracked plate with a bonded patch. It should be emphasized that the theory just described is not tied to the use of the 1D theory of bonded joints: one can calculate the energy change $\delta\mathscr{E}$ using a more elaborate analysis and thereby derive a refined value for $\mathscr{G}_\infty$. However, the 1D theory appears to be sufficiently accurate for preliminary design purposes, and it is extremely convenient to use. For example, one can readily estimate the effect of a variation in the glue-line thickness [8], or of plastic yielding in the adhesive (see below).

Estimates for maximum reinforcement stress and adhesive strain

The reduction of the analysis to that of an overlap joint also leads naturally to upper bounds (and thus to conservative estimates) for other important design parameters besides K_r. For example, it is clear that on going from AA′ to BB′ in Figure 5.5, the reinforcement has to carry an extra stress at $y = 0$ given by $(\sigma_0 t_P)/t_R$, and it is also clear that the maximum stress in the reinforcement, σ^R_{max} occurs at $y = 0$. We can thus derive the following estimate,

$$\sigma^R_{\text{max}} = F/t_R, \tag{5.25}$$

with F given by equation (5.16).

The maximum shear strain in the adhesive will also occur at $y = 0$ provided that the reinforcing patch is tapered around its boundary $\mathscr{D}$, as it is commonly done to relieve the peel stress. That maximum strain can be estimated again by considering the overlap joint which leads to the following upper bound if the adhesive remains elastic,

$$\gamma_{\max}^{A} = \sigma_0 t_P \beta / G_A. \tag{5.26}$$

Within the framework of the one-dimensional theory of joints, one can readily allow for plastic yielding in the adhesive. For an elastic-perfectly plastic adhesive, with a shear yield-stress τ_Y, the adhesive begins to yield when $\sigma_0 t_P \beta = \tau_Y$, and thereafter equation (5.26) should be replaced by

$$\gamma_{\max}^{A} = 0.5\, \tau_Y \{1 + (\sigma_0 t_P \beta/\tau_Y)^2\}/G_A, \tag{5.27}$$

while Λ in equations (5.20) and (5.22) should be replaced by Λ^* given by

$$\pi\Lambda^* = \tfrac{1}{3}(t_A/G_A)E_P t_P \beta (\sigma_0 t_P \beta/\tau_y)^2 \{1 + 2(\sigma_0 t_P \beta/\tau_y)^3\} \tag{5.28}$$

Restoring the overall stiffness.

The principal concern when repairing a cracked plate is to restore the strength to some acceptable level. However, the presence of a crack in a structural member also reduces the overall stiffness of that member, so that it is also of interest to consider whether a bonded repair can be designed so as to restore the original overall stiffness. We first note that when a centre crack is introduced into a plate under load, the change in configurational energy is given by

$$\mathscr{E}_{\text{crack}} = -2\int_0^a \mathscr{G}(a)\mathrm{d}a. \tag{5.29}$$

This energy change is closely related to the change in overall compliance, a relation which is widely used in fracture mechanics to determine the crack extension force $\mathscr{G}(a)$ from the rate of change of compliance with crack extension (see for example [9]). Using this relation in the opposite direction, one can conveniently assess the change in overall compliance after a bonded repair by estimating the change in configurational energy. It is again convenient to proceed in two stages as above.

Consider first the change in configurational energy $\mathscr{E}_{\text{reinf}}$ due to an elliptical patch bonded to an infinite plate under remote biaxial stress. We assume as before that the reinforced region can be treated as an elliptical inclusion of stiffness $E_P t_P + E_R t_R$. Then, using the results given in [1], and the general method of calculating energy variations for inclusion problems [10], we derive

$$\mathscr{E}_{\text{reinf}} = \frac{\pi(\sigma^\infty A)^2}{2E_P(1 - \nu^2)} (S/D)\{C_1(\Sigma - \nu)^2 + C_2(\Sigma - \nu)(1 - \nu\Sigma) + C_3(1 - \nu\Sigma)^2\}, \tag{5.30}$$

$$C_1 = 2(1 + B/A)^2 - 2\nu^2 + S\{(3 + 2\nu - \nu^2)B/A + 2 - 2\nu^2\},$$

$$C_2 = 4\nu(1 + B/A)^2 + (1 + \nu)^2 SB/A,$$

$$C_3 = 2(1 + B/A)^2 + S\{(3 + 2\nu - \nu^2)B/A + 2(1 - \nu^2)(B/A)^2\},$$

and D is given in equation (5.16). The important feature of this result is that, for fixed aspect ratio, the energy change is proportional to A^2, which typically is approximately equal to a^2.

Proceeding next to stage II, the energy change on introducing the crack can be derived from equations (5.29) and (5.24) in the following form,

$$\mathscr{E}_{\text{crack}} = (\sigma_0^2/E_P)(2\pi\Lambda a)\{1 - (\Lambda/a)\log(1 + a/\Lambda)\}. \quad (5.31)$$

Thus $\mathscr{E}_{\text{crack}}$ is proportional to a^2 for $a/\Lambda \ll 1$, and proportional to a for $a/\Lambda \gg 1$.

To minimize the change in overall stiffness we now need to select the patch stiffness and aspect ratio so as to minimize the sum $\mathscr{E}_{\text{reinf}} + \mathscr{E}_{\text{crack}}$. It follows from what we have noted about equations (5.30) and (5.31) that the optimum choice will depend on the value of a/Λ.

5.5 Residual thermal stress due to adhesive curing

The process of adhesive bonding using high-strength structural adhesives generally requires curing the adhesive. This involves heating the reinforced region to a temperature of approximately 120°C, under pressure, for approximately one hour (the precise curing cycle depends on the adhesive being used, see Chapter 6). If the reinforcing patch has a lower coefficient of thermal expansion than the plate being repaired (as it is the case for carbon or boron fibre-composite patches applied to the usual structural metals), a tensile residual stress develops in the plate on cooling after the cure. The magnitude of this stress depends sensitively on the constraint against free expansion at the outer edge of the plate, [11, 2].

To study the effect of this constraint, consider the configuration shown in Figure 5.6, in which an uncracked circular plate of radius R_0 is reinforced by a concentric circular patch of radius R_i. The constraint at the outer edge $r = R_0$ is modelled by a continuous distribution of springs according to the boundary condition

$$\sigma_{rr}^P(r = R_0) = -kE_P u_r^P(r = R_0). \quad (5.32)$$

Now suppose that the inner portion $r \leqq R_i$ is maintained at a temperature T_i during the curing process, while the outer edge $r = R_0$ is maintained at a temperature T_0, with the usual convention that the ambient temperature is taken as the zero of temperature (see for example [12]). The problem is to determine the thermal stress

$$\sigma_0^T \equiv \sigma_{yy}^P(|x| < R_i, y = 0), \quad (5.33)$$

on cooling to ambient temperature after curing.

The solution involves two steps. First we must determine the effect of the edge constraint on the expansion. Using the results given in [12], Section 150, we can derive the following equation for the radial displacement in the plate during the curing process,

$$u_r^P(r = R_i) = \lambda\alpha^P T_i R_i, \quad (5.34)$$

$$\lambda = \frac{(1 + \nu)}{2}\left[1 + \kappa\left\{\frac{T_0}{T_i} + \frac{(T_i - T_0)\{1 - (R_i/R_0)^2\}}{T_i \log(R_0/R_i)^2}\right\}\right]$$

$$\kappa = (1 - \nu)\{1 - (1 + \nu)kR_0\}/[(1 + \nu)\{1 + (1 - \nu)kR_0\}].$$

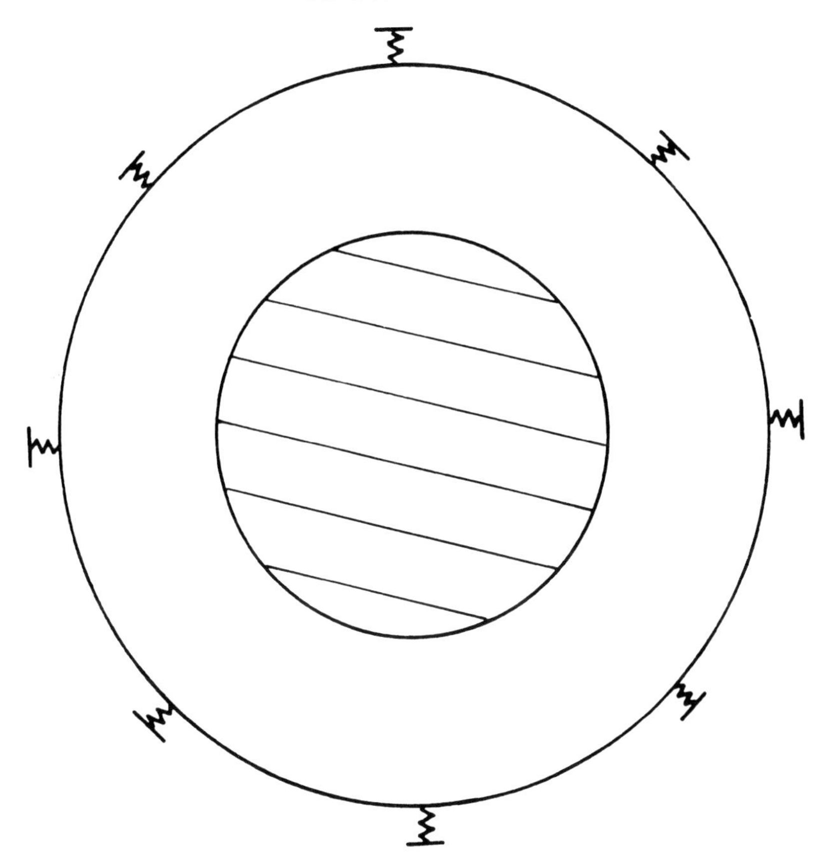

Figure 5.6. Configuration used for calculating the residual thermal stress induced during adhesive curing.

In this equation α^P denotes the linear coefficient of thermal expansion for the plate, and $\lambda\alpha^P$ can be viewed as an *effective coefficient of expansion*. With $k = 0$ and $k \to \infty$ we recover the results for a free edge and a clamped edge respectively [12], k being the spring constant for the constraint according to equation (5.32).

For the second step we now suppose that there is no shear stress in the adhesive during curing, so that the reinforcing patch expands freely, i.e.

$$u_r^R(r = R_i) = \alpha^R T_i R_i, \tag{5.35}$$

and that, on cooling after curing, the bond behaves as a rigid bond ($\beta^{-1} \ll R_i$). One can then show, after some lengthy algebra, that the residual thermal stress is given by

$$\sigma_0^T = \frac{E_P T_i(\lambda\alpha^P - \alpha^R)}{(1 - \nu)[\frac{1}{2}(1 + \nu)\{1 + \kappa(R_i/R_0)^2\} + 1/S]} \tag{5.36}$$

For an *infinite plate*, $R_0/R_i \to \infty$, $kR_0 \to \infty$ and $T_0 \to 0$, so that equation (5.36) reduces to

$$\sigma_0^T \cong \frac{E_P T_i\{(1 + \nu)\alpha^P - 2\alpha^R\}}{(1 - \nu)\{(1 + \nu) + 2/S\}} \tag{5.37}$$

In that limit σ_0^T is insensitive to the precise clamping condition at the outer edge, so that equation (5.37) can be expected to represent an upperbound to the thermal stress which is likely to occur in practice.

At the other extreme of *full reinforcement*, i.e. $R_0 = R_i$, $T_0 = T_i$,

$$\sigma_0^T \cong \frac{E_P T_i\{(1 + \nu)(1 + \kappa)\alpha^P - 2\alpha^R\}}{(1 - \nu)\{(1 + \nu)(1 + \kappa) + 2/S\}}. \tag{5.38}$$

The thermal stress now depends sensitively on the spring constant k, which characterizes the clamping constraint, through its effect on κ which changes from $(1 - \nu)/(1 + \nu)$ for $k = 0$, free-edge condition, to -1 for $k \to \infty$, clamped edge condition.

5.6 Bending effects

So far we have ignored the tendency for out-of-plane bending which would result from bonding a reinforcing patch to only one face of a plate, so that, strictly speaking, the preceding analysis is more appropriate for the case of two-sided reinforcement, with patches bonded to both faces. To assess the bending effects for one-sided reinforcement we shall consider the particular case where the reinforcement is bonded over a strip $|y| \leqq B$, as in Figure 5.7(a). It is again convenient to divide the analysis into two stages.

Stage I: Stress elevation due to bending.
Consider first the effect of one-sided reinforcement on an *uncracked* plate which is subjected to a uniaxial tension

$$\sigma^\infty = \sigma_{yy}^P(y \to \pm\infty).$$

The bending deformation which occurs is indicated in Figure 5.7(b). The normal stress σ_{yy}^P at $y = 0$ will vary across the thickness because of this bending. The problem is to determine the average stress $\bar{\sigma}_0$ in the plate, where

$$\bar{\sigma}_0 \equiv \frac{1}{t_P}\int_0^{t_P} \sigma_{yy}^P(y = 0, z)\mathrm{d}z. \tag{5.39}$$

To do this we shall use the rigid-bond approximation (as in Section 5.3), in conjunction with the conventional theory of cylindrical bending of plates, i.e. we shall assume that the bending deformation of the reinforced portion satisfies the usual kinematic condition that plane sections remain plane. It can then be shown that the neutral surface for the reinforced portion lies at $z = z_0$, where

$$z_0 = \frac{1}{2}\frac{E_P t_P^2 + E_R\{(t_P + t_R)^2 - t_P^2\}}{E_P t_P + E_R t_R}, \tag{5.40}$$

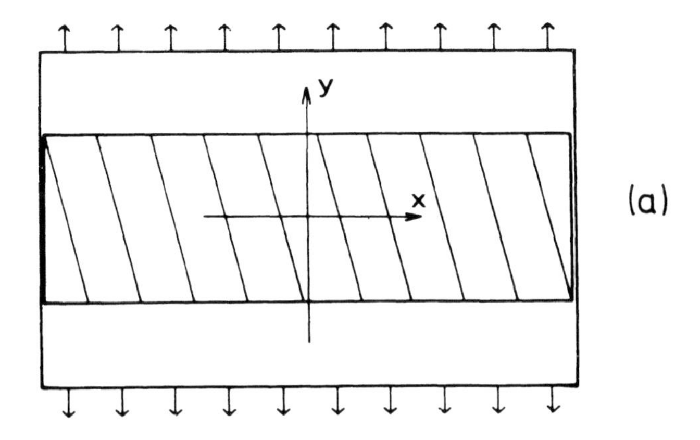

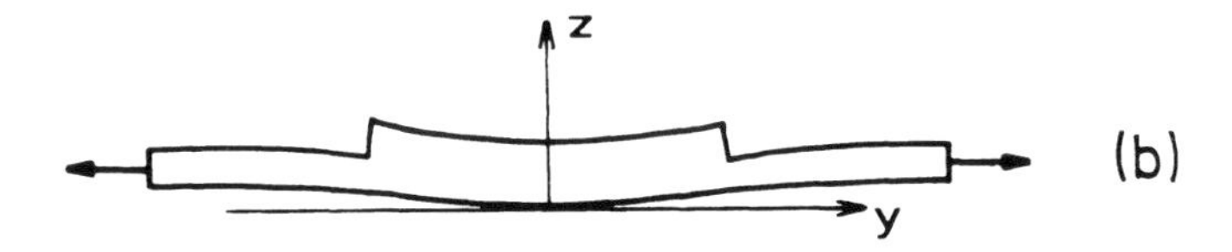

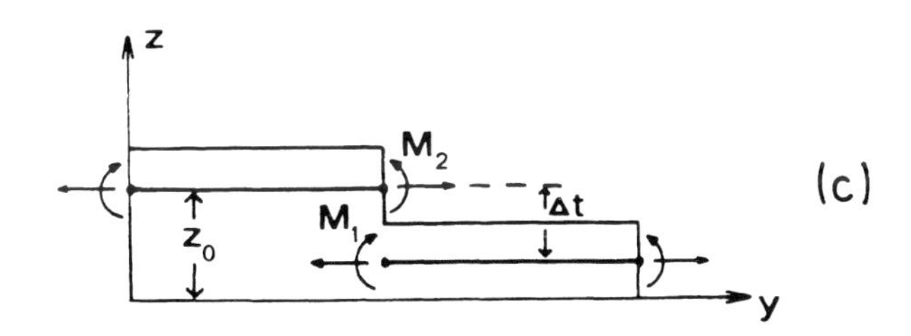

Figure 5.7. Configuration used for assessing out-of-plane bending due to one-sided reinforcement. (a) Plan view; (b) Side view showing nature of bending deformation under load; (c) Sign convention for the bending moments.

with z measured from the un-reinforced face, and the bending stiffness (per unit width along the x-axis) is

$$(EI)_{RP} = \frac{E_P}{3(1 - \nu^2)} \{(t_P - z_0)^3 + z_0^3\} + \frac{E_R}{3(1 - \nu^2)} \{(t_P + t_R - z_0)^3 - (t_P - z_0)^3\} \tag{5.41}$$

Outside the reinforced portion, the neutral surface lies at $z = \frac{1}{2}t_P$ and the bending stiffness is

$$(EI)_P = E_P t_P^3/\{12(1 - \nu^2)\}. \tag{5.42}$$

The configuration of Figure 5.7(b) being symmetrical about $y = 0$, we shall only consider the portion $y \geqq 0$. With the sign convention of Figure 5.7(c), the relation between the bending moment $M(y)$ and the deflection $w(y)$ is

$$M(y) = EIw''(y),$$

where EI is the appropriate bending stiffness. The balance of moments for a segment (y, ∞) then leads to the differential equation

$$EIw''(y) = Fw(y),$$

$$F = \sigma^{\infty} t_P,$$

subject to the boundary conditions

$$w(y \to \infty) = 0, \quad w'(y = 0) = 0.$$

Thus the deflection can be expressed in terms of the (unknown) moments M_1, M_2 at the junction $y = B$, as follows

$$w(B < y < \infty) \equiv w_P(y) = (M_1/F) \exp\{-\chi_P(y - B)\}, \tag{5.43a}$$

$$w(0 \leqq y < B) \equiv w_{RP}(y) = (M_2/F)(\cosh \chi_{RP} y)/(\cosh \chi_{RP} B), \tag{5.43b}$$

$$\chi \equiv \{F/(EI)\}^{1/2}. \tag{5.43c}$$

M_1 and M_2 are now determined by imposing the compatibility conditions [13]

$$w_{RP}(B) - w_P(B) \equiv \Delta t = z_0 - \tfrac{1}{2} t_P, \tag{5.44}$$

$$w'_{RP}(B) = w'_P(B),$$

where Δt denotes the distance between the neutral plane in the reinforced portion $(y < B)$ and the neutral plane in the un-reinforced portion $(y > B)$. This leads to

$$M_1/F = -\Delta t \chi_P/(\chi_P + \chi_{RP} \tanh \chi_{RP} B), \tag{5.45}$$

$$M_2/F = \Delta t \chi_{RP} \tanh \chi_{RP} B/(\chi_P + \chi_{RP} \tanh \chi_{RP} B).$$

The normal stress σ^P_{yy} at $y = 0$ can be viewed as the sum of a direct stress

$$\sigma_0 = \sigma^{\infty}/(1 + S) \tag{5.46}$$

and a bending stress distribution

$$\sigma_{\text{bend}}(z) = E_P(z_0 - z) w''_{RP}(y = 0), \quad 0 < z < t_P,$$

$$= E_R(z_0 - z) w''_{RP}(y = 0), \quad t_P < z < t_P + t_R.$$

The average stress $\bar{\sigma}_0$ defined by equation (5.39) can therefore be expressed in the form

$$\bar{\sigma}_0 = \sigma_0(1 + \omega), \tag{5.47}$$

where the *bending correction* ω is found to be

$$\omega = \{(E_P t_P + E_R t_R)/(EI)_{RP}\}(\Delta t)^2 f(\chi_{RP} B;\, \kappa_P), \tag{5.48}$$
$$f(x; k) = \tanh x/(k \cosh x + \sinh x),$$
$$\kappa_P = \chi_P/\chi_{RP} = \{(EI)_{RP}/(EI)_P\}^{1/2}.$$

Two important features emerge from this analysis. First, bending introduces new characteristic lengths, namely $(\chi_P)^{-1}$, $(\chi_{RP})^{-1}$. By contrast to the load-transfer length β^{-1} defined in Section 5.3, these new characteristic lengths depend not only on the physical parameters of the repaired plate (via the bending stiffnesses $(EI)_P$ and $(EI)_{RP}$), but also on the applied load. Secondly, the function $f(x)$ in equation (5.48) has a maximum in the range of interest. This means that if we view the overlap length $2B$ as the primary variable, all other parameters being kept constant, there is a value of B for which the bending correction is a maximum, but the value of B for which this maximum occurs depends on the applied stress.

These features can be brought out more clearly by considering two special cases, for both of which we shall set

$$\nu_P = \nu_R = \nu = 1/3. \tag{5.49}$$

First, let

$$E_P = E_R = E, \quad t_P = t_R = t. \tag{5.50a}$$

Then one can readily verify that

$$z_0 = t, \quad \Delta t = \tfrac{1}{2}t,$$
$$(EI)_{RP} = 8(EI)_P = 3/4\, Et^3,$$

so that equation (5.48) reduces to

$$\omega = 2/3 f(x; 2\sqrt{2})$$
$$x \equiv \chi_{RP} B = (2B/t)(\sigma^\infty/3E)^{1/2}.$$

The function f for this case has a maximum value of 0.142 when $x = 0.8$, as shown by curve (i) of Figure 5.8, leading to a maximum bending correction

$$\omega_{max} = \tfrac{2}{3} f_{max} = 0.095. \tag{5.51a}$$

So, for example, if we choose

$$\sigma^\infty/E = 2 \times 10^{-3}, \; t = 3\,\text{mm}, \tag{5.52}$$

as representative values for the applied stress and the plate thickness, the bending length would be

$$(\chi_{RP})^{-1} = 5\sqrt{15}t = 58\,\text{mm},$$

and the maximum bending correction would apply for an overlap $2B$ with

$$B = 0.8 \times 58\,\text{mm} = 46\,\text{mm}.$$

As a second illustration, let

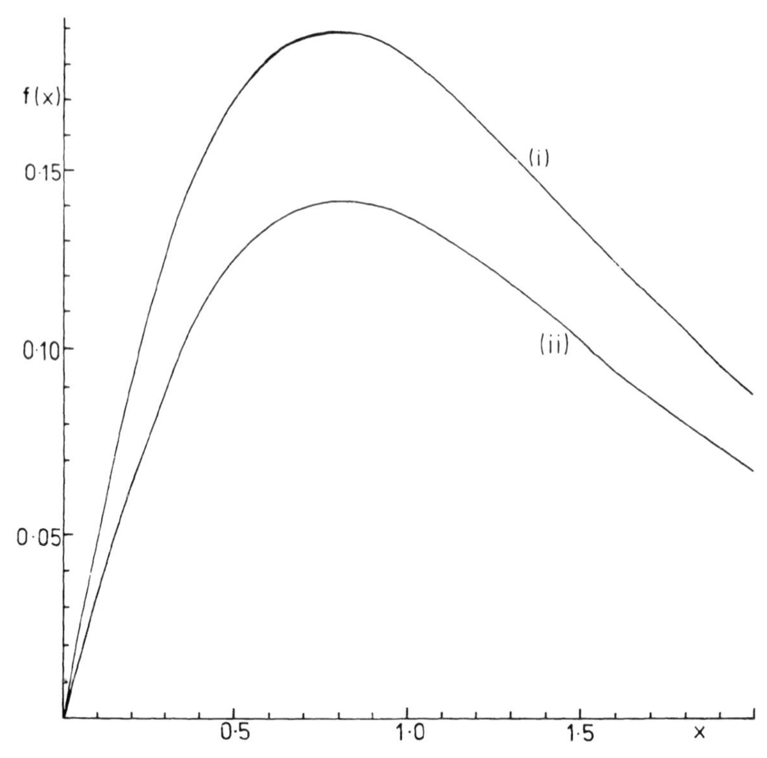

Figure 5.8. Stage I bending correction. The function $f(x;\, k)$ of equation (5.48) for (i) $k = 2\sqrt{2}$, (ii) $k = 1.94$.

$$E_P = E_R/3 = E,\; t_P = 3t_R = t. \tag{5.50b}$$

Then one finds that

$$z_0 = 5t/6, \quad \Delta t = t/3, \quad (EI)_{RP} = \frac{17}{48}Et^3$$

$$\kappa_P = \sqrt{34}/3 = 1.94,$$

so that

$$\omega = \frac{32}{51} f(x;\, 1.94),$$

$$x \equiv \chi_{RP}B = (4B/t)(3\sigma^\infty/17E)^{1/2}.$$

For this case f has a maximum value of 0.19 when $x = 0.78$, as shown by curve (ii) of Figure 5.8. Using the values given in equation (5.52), we now find that the bending length is

$$(\chi_{RP})^{-1} = 13.3t = 40\,\text{mm},$$

and that the maximum bending correction

$$\omega_{\max} = (32/51)\, f_{\max} = 0.119, \tag{5.51b}$$

is required when the overlap length is

$$B = 0.78 \times 40\,\text{mm} = 31\,\text{mm}.$$

In practice one would generally try to avoid using overlap lengths which lead to the maximum bending correction. The preceding examples suggest, however, that the maximum bending correction at stage I is relatively small, being typically less than 12%.

Stage II: Asymptote for $\mathcal{G}$ and K.
Repeating the argument used in Section 5.4, we can again derive the crack extension force for a semi-infinite crack from the energy change (per unit width along the x-axis) on going from configuration (a) of Figure 5.2 to configuration (d), but allowing now for out-of-plane bending. For that purpose we shall treat configuration (d) as being equivalent to configuration (e) of Figure 5.2, so that effectively the same analysis that was used above for stage I can now be used to calculate the induced bending moment $M(y)$. The bending energy can then be obtained by integrating $M^2/(EI)$.

To bring out more clearly the essential features we shall assume an infinite overlap ($B \to \infty$) as in Section 5.4. Then, the only characteristic lengths involved are the length H shown in Figure 5.2(e) and the bending lengths $(\chi_R)^{-1}$, $(\chi_{RP})^{-1}$. The crack extension force can be written in the form (cf. equation (5.20)).

$$\mathcal{G}_\infty = (\sigma_0^2/E_P)\pi\Lambda(1 + \Omega), \tag{5.53a}$$

where the bending correction is found to be

$$\Omega = \tfrac{1}{2}\{S(E_P/\sigma_0)(\Delta t)^2\beta\chi_R\}g(\chi_R H;\, \kappa_R), \tag{5.54}$$

$$\chi_R = \{(1 + S)\sigma_0 t_P/(EI)_R\}^{1/2},$$

$$g(x;\, k) = (x + \sinh x \cosh x + k \sinh^2 x)/(\cosh x + k \sinh x)^2,$$

$$\kappa_R \equiv \chi_R/\chi_{RP} = \{(EI)_{RP}/(EI)_R\}^{1/2}.$$

To simplify this result, let us consider first the special case of equations (5.49) and (5.50a). Then equation (5.54) reduces to

$$\Omega = (E/3\sigma_0)^{1/2}\beta t g(x;\, 2\sqrt{2}), \tag{5.55a}$$

$$x = \chi_R H = (\sigma_0/3E)^{1/2}(8H/t),$$

where g for this case is shown as curve (i) in Figure 5.9. Consider next the special case of equations (5.49) and (5.50b), which is representative of boron fibre reinforcement bonded to an aluminium plate. Then

$$\Omega = (4/9)(3E/\sigma_0)^{1/2}\beta t g(x;\, \sqrt{34}), \tag{5.55b}$$

$$x = \chi_R H = (3\sigma_0/E)^{1/2}(8H/t),$$

where g is shown as curve (ii) of Figure 5.9. These simplified forms for the stage II bending correction show clearly the important non-dimensional parameters, which are (i) the load factor σ_0/E, (ii) the ratio of thickness to load-transfer length

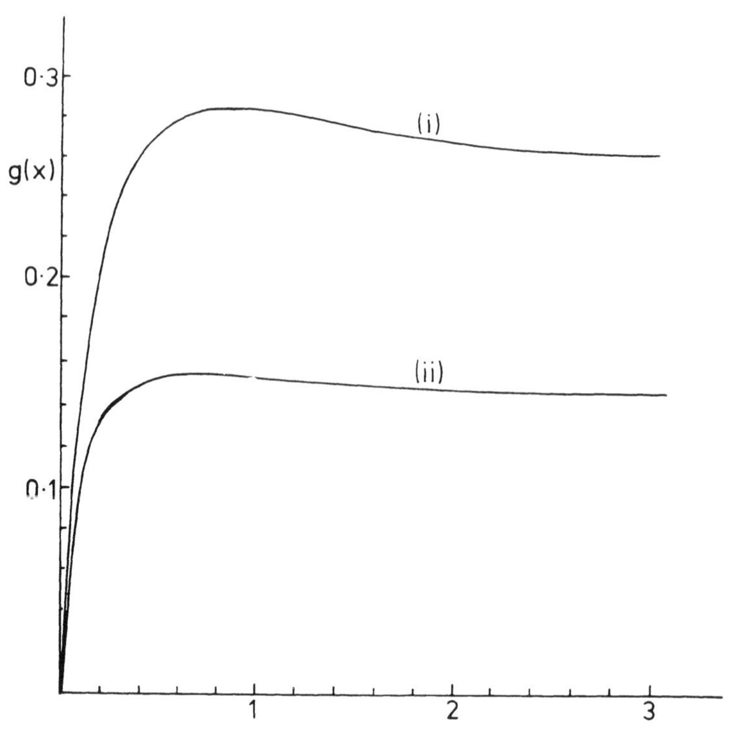

Figure 5.9. Stage II bending correction. The function $g(x; k)$ of equation (5.54), for (i) $k = 2\sqrt{2}$, (ii) $k = \sqrt{34}$.

βt, and (iii) the ratio t/H. So far we have not specified the length H, but the argument in Section 5.2 suggests that H should be equal to β^{-1}, at least as a reasonable first approximation.

Using as representative values

$$\sigma_0/E = 10^{-3}, \quad t = 3\,\text{mm}, \quad H = \beta^{-1} = 5\,\text{mm}, \tag{5.56a}$$

we find from equation (5.55a), for the case of equal moduli of plate and reinforcement, that

$$\Omega = 10.95\,g(0.24; 2\sqrt{2}) = 2.41, \tag{5.56b}$$

while, for the case of unequal moduli, equation (5.55b) leads to

$$\Omega = 14.61\,g(0.73; \sqrt{34}) = 2.28. \tag{5.56c}$$

Thus, by contrast to the stage I bending correction ω, which was typically less than 12%, the stage II bending correction can be quite substantial, amounting in the above cases to an increase in $\mathscr{G}_\infty$ by a factor of 3.3 ($= 1 + \Omega$), or an increase in the corresponding stress intensity factor by a factor of 1.8, the limiting value of K_r being given by

$$\bar{K}_\infty = \sigma_0\{\pi\Lambda(1 + \Omega)\}^{1/2}, \tag{5.53b}$$

instead of equation (5.22).

5.7 Partial reinforcement

In practice it is not always possible for the reinforcement to cover completely the crack, as it has been assumed so far. The aim of this section is to describe an approach which greatly simplifies the task of assessing the efficiency of reinforcements which only partially cover the crack.

Consider the configuration shown in Figure 5.10(a). As before, the analysis can be divided into two stages. At stage I, it is assumed that the plate is uncracked, and the problem is to determine the stress at the prospective location of the crack,

$$\sigma_0(x) = \sigma_{yy}^P(|x| \leqq a, y = 0). \tag{5.57}$$

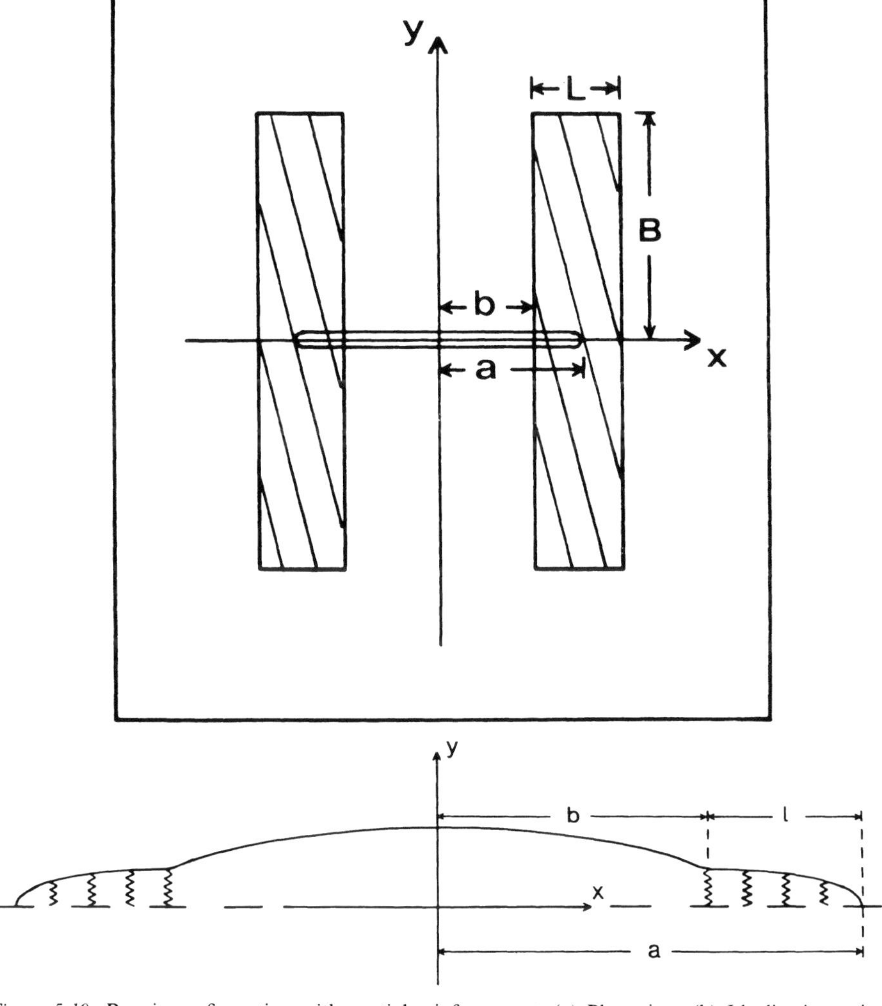

Figure 5.10. Repair configuration with partial reinforcement. (a) Plan view. (b) Idealization using distributed springs for stage II analysis.

To that end, we can again use the inclusion analogy described in Section 5.4, but now the stress analysis must be done numerically as there does not seem to be analytical solutions for problems involving two inclusions.

The next step is to cut the plate along the line segment $|x| \leqq a$, $y = 0$, and to apply tractions to the crack faces so as to cancel the stress $\sigma_0(x)$ determined at stage I. The essential reinforcing action at this second stage is the restraint on the crack opening due to the bonded reinforcements. The basic simplification which is proposed here is that this restraining action can be represented by a continuous distribution of springs acting between the crack faces, as illustrated in Figure 5.10(b). This idealization reduces the problem at stage II to two parts: (i) determine the appropriate constitutive relation (i.e. stress-displacement relation) for the springs, and (ii) solve a one-dimensional integro-differential equation for the crack opening

$$\begin{aligned}\delta(x) &= u_y^P(x, y \to 0+) - u_y^P(x, y \to 0-), \quad |x| < a. \\ &= 2u_y^P(x, y \to 0+), \quad |x| < a. \end{aligned} \tag{5.58}$$

To illustrate the procedure we shall assume that the reinforcements satisfy the condition $L \ll B$ (with L, B, as shown in Figure 5.10(a)), so that, as a reasonable first approximation, the stress $\sigma_0(x)$ at the prospective crack location can be taken to be equal to the applied stress σ_{yy}^{∞} (cf. equation (5.2)). We shall also assume that the equivalent springs are linear so that the stress exerted by the springs on the crack faces can be written in the form

$$\begin{aligned}\sigma^S(x) &= \tfrac{1}{2}kE_P u_y^P(x, y \to 0+), \quad b < |x| < a, \\ &= 0, \quad |x| < b,\end{aligned} \tag{5.59}$$

where k is the normalized spring constant. Finally, we shall assume that the appropriate value of k can be determined from the stress-displacement relation for the overlap joint, as given by equation (5.14) using the 1-D theory. This leads to the following expression for k,

$$k = 2/(\pi\Lambda), \tag{5.60}$$

where Λ is the characteristic length defined in equation (5.21). With these assumptions, the problem of determining the crack opening $\delta(x)$ can be reduced to that of solving the following integro-differential equation,

$$\frac{1}{2\pi}\int_{-a}^{a} \frac{\mathscr{D}(t)}{x - t}\,\mathrm{d}t = -(2/E_P)[\sigma_0(x) - \sigma^S(x)], \quad |x| < a, \tag{5.61}$$

where $\mathscr{D}(x)$ denotes the *dislocation density*, which is the derivative of the crack opening $\delta(x)$ and thus is an odd function of x,

$$\delta(x) = \int_x^a \mathscr{D}(t)\,\mathrm{d}t \tag{5.62}$$

$$\mathscr{D}(x) = -\delta'(x) = -\mathscr{D}(-x). \tag{5.63}$$

In the absence of springs (or with $k = 0$), equation (5.61) would reduce to the familiar singular integral equation for the dislocation density in a centre crack, which can be solved analytically [14, 15]. With springs present however, it does not seem possible to derive an analytical solution (cf. [15], chapter 17), but it is relatively simple to solve the equation (5.61) numerically. For that purpose it is convenient to use the following change of variables

$$x = a \sin \theta,\ t = a \sin \phi,\ b = a \sin \zeta, \tag{5.64}$$

$$\mathscr{D}(x) = (4\sigma_0/E_P) \sec \theta V(\theta),$$

so that equation (5.61) can be reduced to the following form, on using equation (5.63),

$$\frac{4}{\pi} \int_0^{\pi/2} \frac{V(\phi) \sin \phi}{\cos 2\theta - \cos 2\phi} \mathrm{d}\phi = 1 - ka \int_\zeta^{\pi/2} V(\phi)\mathrm{d}\phi,\ \zeta \leqq \theta < \pi/2,$$

$$= 1,\ 0 < \theta < \zeta. \tag{5.65}$$

This equation can be solved straight-forwardly by using the trapezoidal rule to replace the integrals by sums, with mesh refinement around the point $\theta = \zeta$. The change of variables in equation (5.64) gives effectively a built-in mesh refinement near the crack tip at $\theta = \pi/2$.

Estimates for the variables which are of interest in the context of crack patching can be derived from the dislocation density $\mathscr{D}(x)$, and therefore, from $V(\theta)$, as follows:

(i) *The stress intensity factor*

$$K_r = \lim_{x \to a} \{\pi(a - x)/8\}^{1/2} E_P \mathscr{D}(x),$$

$$= K_0 V(\theta = \pi/2), \tag{5.66}$$

with

$$K_o = \sigma_0(\pi a)^{1/2}.$$

(iii) *The maximum adhesive shear strain*

$$\gamma^A_{max} = [u_p(x = \pm b, y \to 0+)]/t_A,$$

$$= [2\sigma_0 a/(E_p t_A)] \int_\zeta^{\pi/2} V(\phi)\mathrm{d}\phi. \tag{5.67}$$

The solution $V(\theta)$ of equation (5.65) depends on two non-dimensional parameters, namely ka and b/a (= $\sin \zeta$). It is therefore appropriate to present the normalized estimates for K_r and γ^A_{max} in the form of graphs for

$$F(ka; b/a) \equiv V(\theta = \pi/2), \tag{5.68}$$

and for

$$f(ka; b/a) \equiv \int_\zeta^{\pi/2} V(\phi)\mathrm{d}\phi, \tag{5.69}$$

as shown in Figures 5.11 and 5.12 respectively. Some asymptotic properties of these functions can be derived analytically and will be presented elsewhere [16].

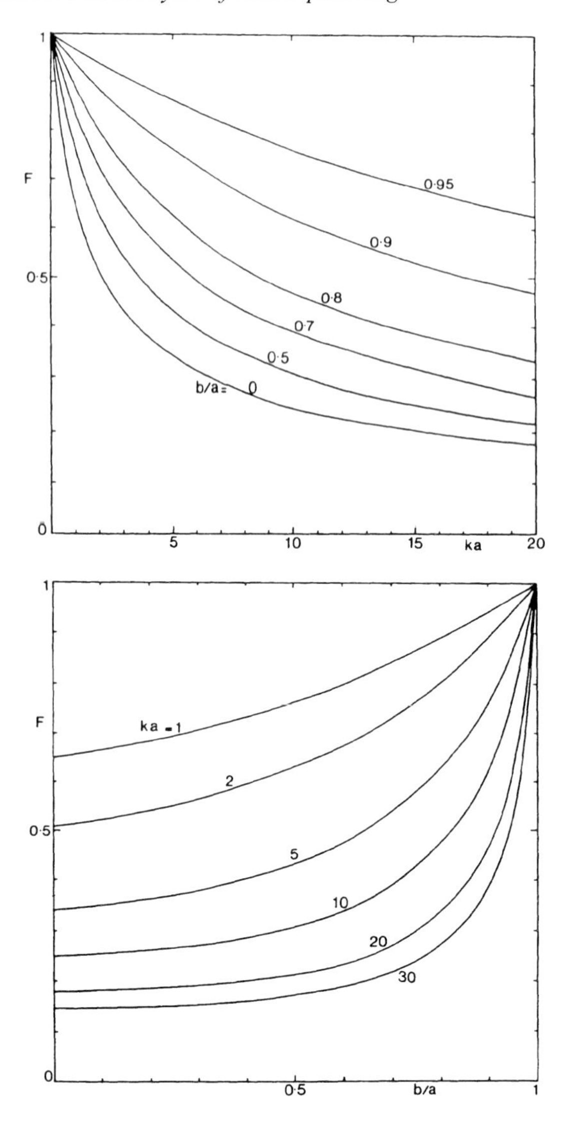

Figure 5.11. The normalized stress intensity factor for the sprung crack.

As an application of the preceding results, consider the problem of estimating the stress intensity factor K as a function of crack length for the configuration of Figure 10(a), in which we now view the spacing $2b$ between the reinforcements as being fixed. (This problem is relevant to the experiment reported in Section 6.6.3). For $a < b$, we can ignore altogether the presence of the reinforcements (provided $L \ll B$, which we shall assume to be the case, and ignoring out-of-plane bending), so that

$$K = K_o = \sigma_0(\pi a)^{1/2}, \quad \text{for } a < b.$$

For $a > b$, the configuration can be viewed as one of partial reinforcement, so that

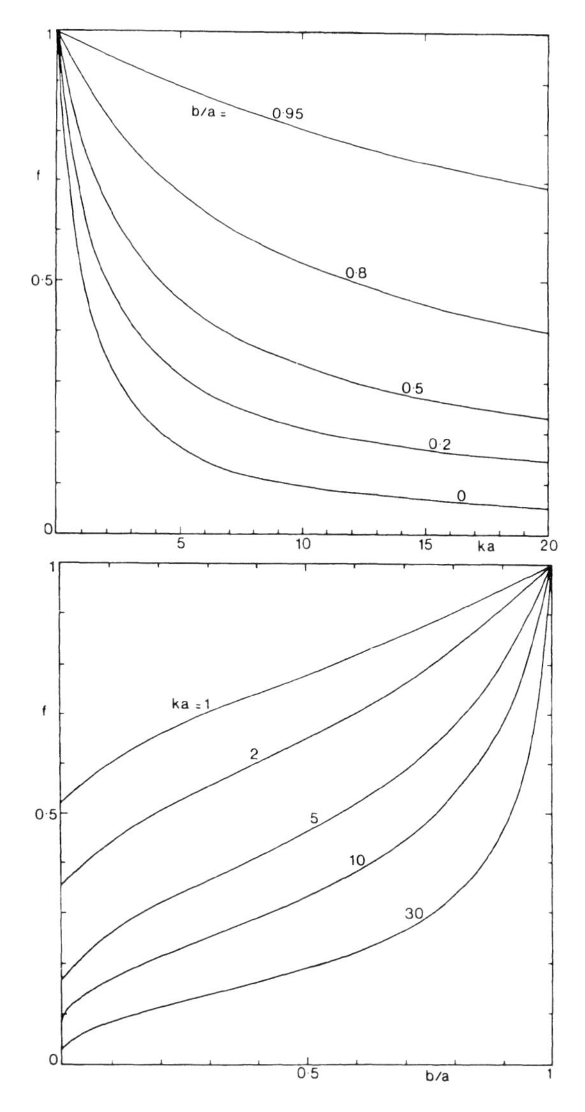

Figure 5.12. The normalized maximum spring-stretch.

we can estimate K from the results given in Figure 5.11 as follows. It is first necessary to specify a value for kb, i.e. for the width of the un-reinforced gap relative to the intrinsic length k^{-1} which is characteristic of the reinforcement. Let us choose $kb = 10$, which is the appropriate value for the experiment in Section 6.6.3. Then, from the curve for $b/a = 0.9$ in Figure 5.11, we can derive the stress intensity factor when

$$ka = kb/0.9.$$

Thus, with $kb = 10$, the relevant point on the curve for $b/a = 0.9$ is that corresponding to $ka = 10/0.9 = 11.11$, for which $F = 0.6$. We therefore have

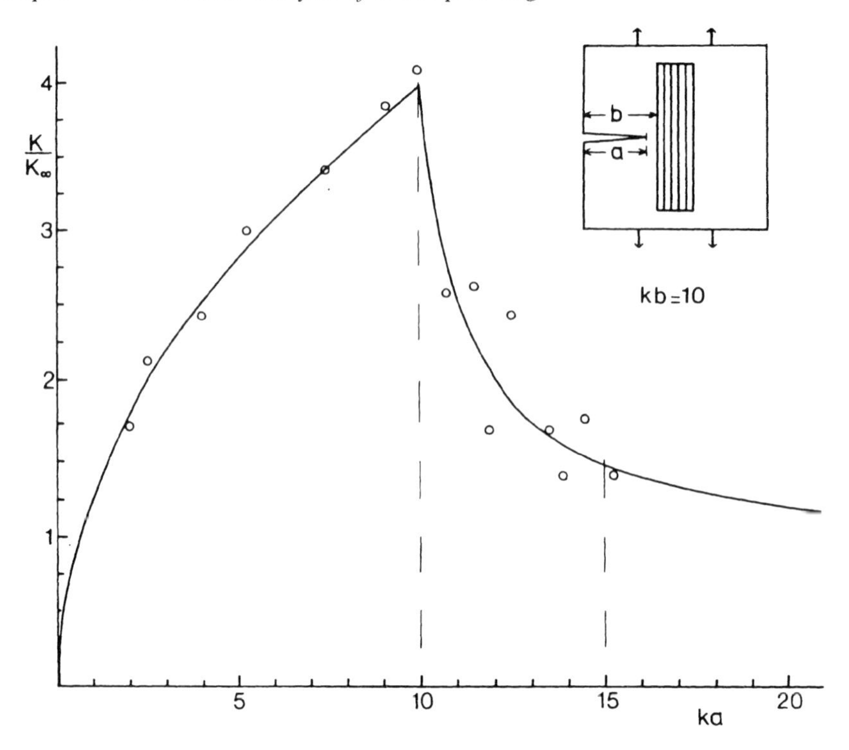

Figure 5.13. Theoretical variation of K with crack length for partial reinforcement with $kb = 10$. Open circles are experimental values for K (Baker 1985) derived from fatigue-crack growth rate using the configuration shown inset, for which the reinforcement covered the crack tip over the range $10 \leqslant ka \leqslant 15$.

$$K(ka = 11.11) = 0.6\sigma_0(\pi a)^{1/2}.$$

Similarly we can derive estimates of K for increasing values of ka from the remaining curves for decreasing values of b/a. The resulting variation of K with crack length is shown in Figure 5.13, where we have normalized K relative to the limiting value

$$K_\infty = \sigma(2/k)^{1/2}$$

which would be derived from equations (5.22) and (5.60). Thus Figure 5.13 in fact shows the variation of

$$K/K_\infty = (\pi ka/2)^{1/2}, \quad a \leqq b,$$
$$= (\pi ka/2)^{1/2}\, F(ka;\, b/a),\ a > b.$$

The good agreement between the theoretical curve in Figure 5.13 and the experimental measurements (shown as open circles) indicates that replacing the reinforcement by distributed springs is indeed sufficiently accurate to be useful in practice.

5.8 Conclusion

It has been shown that the efficiency of crack-patching can be conveniently assessed by proceeding in two stages, because the load transfer involved at each stage can occur more-or-less independently of the other. Consequently, the relevant material and configurational parameters, and hence the appropriate simplifying assumptions, may be different for each of these two stages. At stage I it is assumed that the reinforcement is bonded to an uncracked component, and the problem is to determine the resulting redistribution of stress at the prospective crack location. The important load transfer at this stage generally occurs around the boundary of the reinforcement. Consequently this load transfer depends primarily on the geometrical characteristics of the repair (for example, the aspect ratio of the bonded patch) and does not depend sensitively on the adhesive parameters. At stage II on the other hand, the load transfer generally occurs in the immediate vicinity of the crack, so that the precise shape and dimensions of the patch and the cracked component can be largely irrelevant.

By using this approach we have derived analytical estimates for the design parameters which are of primary interest in crack patching, for the case of full coverage of the crack by the reinforcement. For partial reinforcement, we have simplified stage II of the analysis by using continuously distributed springs, acting between the crack faces, to represent the restraint of the reinforcement on the opening of the crack. An important feature of this modelling procedure is again that it clearly separates the two aspects involved. The first is to determine the appropriate constitutive relation for the springs from the behaviour of an overlap joint corresponding to the repair. The second is to solve an elastic crack problem with displacement-dependent internal loading on the crack faces. These conceptual divisions of the mechanics of bonded reinforcements have proved useful not only in providing explicit estimates of the important design parameters but also in simplifying the experimental measurement of the relevant material properties. This simplification is especially valuable when fatigue or environmentally-induced damage can occur in the adhesive, because it is only necessary to characterize the behaviour of an overlap joint made up of the same materials as in the repair. These aspects of adhesive mechanical properties are discussed in Chapter 6.

References

[1] Rose, L.R.F., *Int. J. Solids Structures*, 17, 827–838 (1981).
[2] Rose, L.R.F., *Int. J. Fracture*, 18, 135–144 (1982).
[3] Sneddon, I.N., in *Adhesion*, D.D. Eley (editor). Oxford University Press (1961).
[4] Rose, L.R.F., *Math. Proc. Camb. Phil. Soc.*, 92, 351–362 (1982).
[5] Rose, L.R.F. and de Hoog, F.R., *Q.J. Mech. App. Math.*, 36, 419–436 (1983).
[6] Ratwani, M.M., *J. Engng. Materials and Technology*, 100, 46–51 (1978).
[7] Ratwani, M.M., *AIAA Journal*, 17, 988–994 (1979).
[8] Rose, L.R.F., *J. Adhesion*, 14, 93–103 (1982).
[9] Knott, J.F., *Fundamentals of Fracture Mechanics*, Section 4.4. Butterworths, London (1973).
[10] Eshelby, J.D., *Proc. R. Soc. Lond.*, A241, 376–396 (1957).

[11] Jones, R. and Callinan, R.J., *J. Structural Mech.*, 8, 143–149 (1980).
[12] Timoshenko, S. and Goodier, J.N., *Theory of Elasticity*, 3rd ed., Chap. 13. McGraw-Hill, NY (1970).
[13] Fraser, W.B., *Int. J. Solids Structures*, 11, 1245–1256 (1975).
[14] Bilby, B.A. and Eshelby J.D., in *Fracture*, Vol. 1, H. Liebowitz (editor). Academic Press, NY (1968).
[15] Muskhelishvili, N.I., *Singular Integral Equations*, Noordhoff, Groningen (1953).
[16] Rose, L.R.F., *J. Mech. Phys. Solids* (in press 1987).

A.A. BAKER

6

Crack patching: experimental studies, practical applications

6.1 Introduction

Adhesively bonded patches of advanced fibre composites, such as boron/epoxy, can provide highly efficient and cost-effective repairs for cracked metallic aircraft components. This approach, known in Australia as crack-patching, pioneered by Aeronautical Research Laboratories since about 1972, has been successfully used to repair Royal Australian Air Force aircraft, including Hercules, Mirage and F111, suffering from cracking due to fatigue or to stress-corrosion.

This chapter mainly presents an overview of our experimental work on this topic, together with a preliminary design approach for estimating the minimum thickness patch to be employed in a given repair situation. Experimental aspects discussed include adhesive fatigue; influence of adhesive cure temperature; surface-treatments for repair bonding; residual-stress effects due to patching; studies on overlap joints, simulating repairs and crack propagation behaviour in patched panels. Finally, some practical applications of crack-patching are briefly described; greater details of practical applications are provided in references [1] to [4].

The advantages and disadvantages of crack-patching repairs relative to traditional procedures based on mechanically fastened metallic patches are best considered under two headings: (i) adhesively bonded reinforcement versus mechanically fastened repairs, and (ii) composites versus metal for bonded repairs.

Adhesively bonded repairs

Compared to mechanical methods such as riveting or bolting, adhesive bonding provides more uniform and efficient load transfer into the patch from the cracked component and introduces much lower stress concentrations. Adhesive bonding produces a sealed interface, thereby reducing the danger of crevice corrosion and leakage (e.g. from a fuel tank). Most importantly, patches can be applied (and if required, removed) with little damage to the parent structure. In contrast, methods based on mechanical fastening can result in considerable damage, for example:

(i) additional fastener holes introduce stress-concentrations, which can encourage fatigue cracking – poorly drilled holes (difficult to avoid in in-situ repairs) are particularly bad in this respect,

(ii) in-situ drilling may cause internal damage to items such as hydraulic lines and electrical wiring, and introduce swarf into the structure,

(iii) poor mechanical fastening procedures may result in fretting damage, encouraging fatigue cracking or high residual stresses, encouraging stress-corrosion cracking, and

(iv) loss of rivets on an outside surface repair, due to mechanical working of the fastened region, may result in costly FOD to engines.

The main disadvantages of adhesive bonding in repair applications are (a) relatively large load transfer lengths may be required, (b) special preparation of surfaces is required prior to bonding, and (c) heat and pressure may be needed to cure the adhesive.

Composite reinforcement

Some of the advantages of unidirectional boron/epoxy (b/ep) and graphite/epoxy (gr/ep) composites for patch materials follows from their mechanical and physical properties, given in Table 6.1, where they are compared with typical aircraft grade aluminium and titanium alloys.

Advantages of composite patches include (i) high directional stiffness which allows use of comparatively thin patches (important for external repairs) and allows reinforcement to be applied specifically in desired directions, (ii) high failure strain and durability under cyclic loading, which minimises the danger of patch failure, (iii) low density, an important advantage where changes in the balance of

Table 6.1. Relevant material mechanical and physical properties.

Material	Modulus GPa	Shear modulus GPa	Critical strain $\times 10^{-3}$	Specific gravity	Thermal expansion coefficient $(^{\circ}C)^{-1} \times 10^{-6}$
b/ep (unidirectional)	208 max 20 min	7	7.3	2.0	4.5 min 23 max
gr/ep (unidirectional)	130 max 12 min	5	13–15	1.6	0.4 min 28 max
Aluminium alloy (7075 T6)	72	27	6.5	2.8	23
Aluminium alloy (2024 T3)	72	27	4.5	2.8	23
Titanium alloy (6 Al/4V)	110	41	8.0	4.5	9

Notes: (a) Maximum modulus and minimum expansion coefficient are in the fibre direction, other values given are for the transverse direction. (b) Shear modulus values given for the composites are for in-plane deformation. (c) Critical strain refers to failing strain in tension for the composites and yield strain for the metals.

a component must be avoided, as for example in the repair of control surfaces, and (iv) excellent formability, allowing low-cost manufacture of patches with complex contours.

Surface-treatment of the composite patches for durable adhesive bonding is much less of a problem than for metals, since surface abrasion is all that is required. Alternatively, the composite patch can be cocured on the cracked component with the adhesive, simplifying the patch fabrication procedure and also avoiding the need for any surface preparation of the patch.

In most repair applications use of unidirectional patches (all 0° plies) will provide the highest reinforcement efficiency. The unidirectional composite generally has sufficient strain capacity to survive biaxial straining of the metal substrate without splitting. However, in some applications with high biaxial tension, it may be desirable to provide shear and transverse reinforcement. This can be achieved by incorporating a small proportion of plies oriented at $\pm 45°$ and 90°.

The main disadvantage in employing composite patches results from the mismatch in thermal expansion coefficient between the composite and the metal, Table 6.1. When adhesives curing at elevated temperature are used to bond the patch, the metal can suffer a tensile residual stress which may adversely influence its performance in several ways, as discussed later.

Both b/ep and gr/ep have the above advantages and disadvantages for use as patch materials. However, b/ep is considered to be the better choice because of its (a) superior combination of strength and stiffness, (b) higher expansion coefficient, which reduces the severity of the residual stress problem, (c) low electrical conductivity, which avoids the danger associated with gr/ep of inducing galvanic corrosion of the metal; the low conductivity of b/ep allows use of simple eddy current procedures to detect and monitor cracks under the patch. However, gr/ep would be preferable if patches with low radii of curvature were required (less than 30 mm) or if b/ep availability or cost was an important consideration.

6.2 Adhesive system and process selection

Success in crack-patching depends largely on the correct selection of adhesive system and surface-treatment procedure. The basic requirement is for an adhesive system which, although applied under field conditions, can withstand the stresses, temperature and chemical environments encountered in aircraft service. The durability required of a repair is dependent on intended permanence; however, for the repair to perform at all usefully the adhesive system must be capable of providing, at least temporarily, satisfactory load transfer under operating conditions. To achieve and maintain load transfer it is most important that the adhesive and adhesive/adherend interface are resistant to failure under cyclic loading due to cyclic shear and peeling stresses and the adhesive is resistant to stress-relaxation.

6.2.1 Fatigue and stress-relaxation studies

A simple low-cost procedure was developed, at an early stage in the patching technology, to allow evaluation of fatigue and comparative stress-relaxation

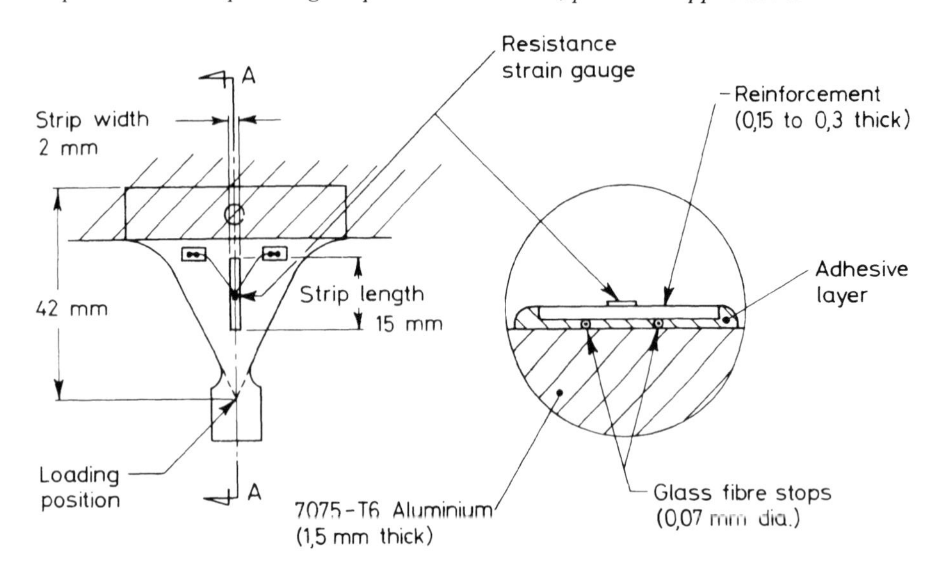

Figure 6.1. Schematic illustration of the constant stress cantilever specimen.

properties of candidate adhesive systems. The procedure [8] is based on measurements of the efficiency of strain transfer from a loaded metal substrate to a composite reinforcement adhesively bonded to the metal surface.

The specimen employed, depicted in Figure 6.1 is essentially a constant stress cantilever beam to which the reinforcement is bonded with the candidate adhesive.

The efficiency of strain transfer to the reinforcement is a function of (i) thickness and shear modulus of the adhesive layer, (ii) substrate modulus, and (iii) length and thickness of the reinforcing strip. In a given specimen all of the above parameters remain constant except that (a) the effective patch length decreases if a fatigue crack grows in the adhesive layer – usually from the high shear stress region at each end of the patch, or (b) the effective shear modulus of the adhesive decreases due to stress-relaxation. These two possibilities are illustrated in Figure 6.2(a) and (b) which depict the relation between (i) the ratio of e_R/e_P, strain in the reinforcement to strain in the metallic component, and (ii) the length of the reinforcement. Thus changes in the measured strain in the strain gauge mounted on the reinforcement can be used to obtain estimates of cracking or relaxation in the adhesive.

Stress-relaxation measurements were made by loading the cantilever specimen to a given level of surface strain for a prescribed time t and then unloading and measuring the resulting compression strain e_u in the reinforcement. The ratio e_u/e_0, compressive strain to the initial strain in the reinforcement, was taken as an indication of the degree of relaxation in time t. Measurements were made on specimens with b/ep and gr/ep reinforcements (studies for comparison were also made on specimens with boron fibre reinforced aluminium and 7075 T6 reinforcements), with several candidate adhesives. Some of the results obtained for the gr/ep reinforced specimens are listed in Table 6.2 where the inferior performance of the polyamide cured epoxy is evident.

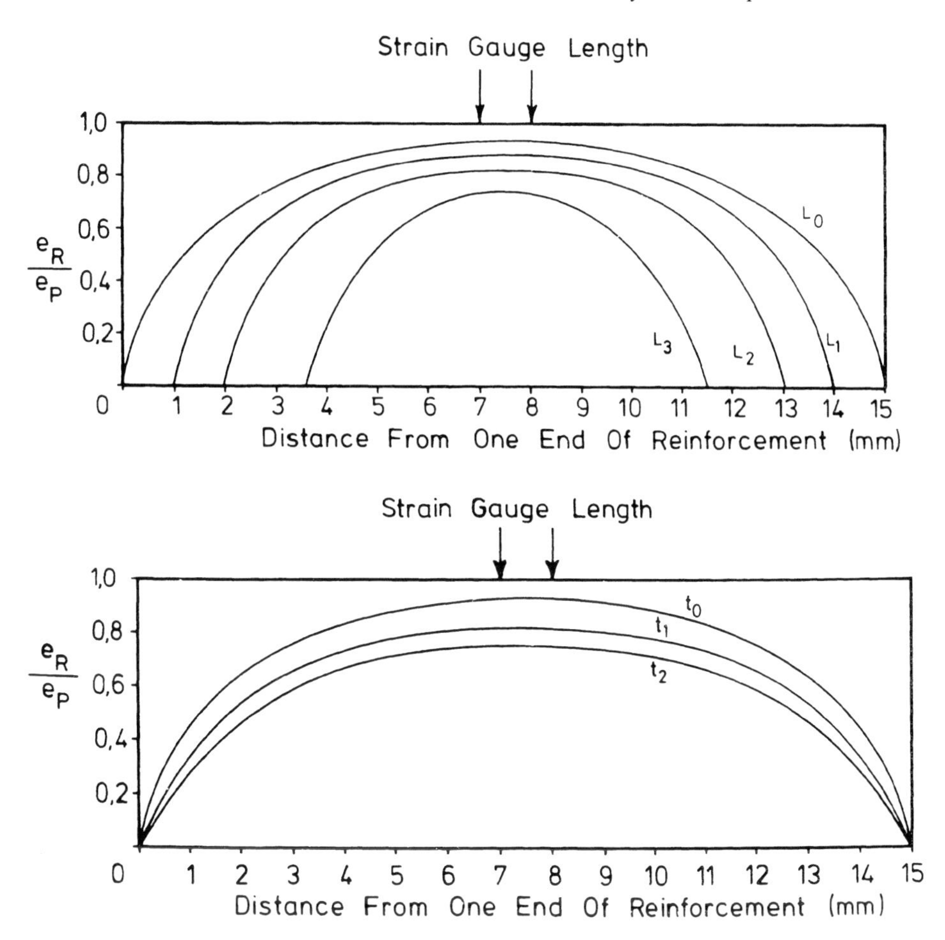

Figure 6.2. Schematic illustrations of the strain distribution in the reinforcement following (a) cracking in the adhesive from each end of the reinforcement, and (b) relaxation for increasing times t_0, t_1, t_2.

Similar tests to the above were also performed at elevated temperatures and after exposing the patch system to moisture. It was concluded from these tests that the two most promising adhesives, the epoxy/amine and the epoxy-nitrile, (AF126 by 3M Co) would only be useful up to about 80°C.

The influence of cyclic loading on the adhesive system was studied by subjecting the cantilever specimens to loading in reversed bending at 50 Hz, a frequency too high for any significant stress-relaxation effects to occur. Above a certain level of strain in the specimen, the strain in the reinforcement was observed to decrease with increasing numbers of cycles, N. Plots of $\frac{1}{2}e_s$ (the peak strain) versus N (Figure 6.3), show this effect for the adhesives studied. This indicates that epoxy/amine and the AF126 adhesives once again have superior properties to the other adhesives studied.

In order to study the nature of the fatigue damage in the adhesive layer, taper sections were made through the reinforcement region of the various specimens

Table 6.2. Relative strain-relaxation measurements for the candidate adhesives after five minute loading under ambient conditions. The surface strain on the cantilever specimen was about 1450 $\mu\varepsilon$, the reinforcement was gr/ep, 0.3 mm thick.

Adhesive type	Cure temp °C/ time h	Thickness mm	e_u/e_0
Ethyl-cyanoacrylate	40/12	0.07	0.02
Epoxy/polyamide	90/19	0.1	0.40
Epoxy/amine	100/19	0.05	0.00
Epoxy/nitrile	125/4	0.09	0.01

Notes: (a) Surface-treatment of the metal was chromic acid FPL etch followed by phosphoric acid tank anodise. (b) Surface-treatment of the composite was by alumina grit-blast.

tested. Typical fatigue damage for each type of adhesive is shown in Figure 6.4, the cracking found provides justification for the assumption that decreases in e_s can be interpreted as cracking in the adhesive layer. It is interesting to note that the cracking usually occurs in $\pm 45°$ principal stress planes for adhesives without polymer reinforcing fibres. The Dacron polymer support fibres in the AF126 adhesive appear to interact strongly with the cracking, probably partially as a result of the relatively weak fibre/adhesive interface. It can also be seen that the epoxy/amine adhesive is prone to cracking (or disbonding) close to the adhesive/metal interface; this adhesive was generally found to be very sensitive to surface-treatment of the metal.

If the assumption is made that the total decrease in e_s results from fatigue cracking in the adhesive, it is possible to estimate the rate of crack propagation by estimating the effective length of the reinforcement strip as a function of N. The effective length can be obtained [8] from the ratio of strain e_s in the strip after cycles

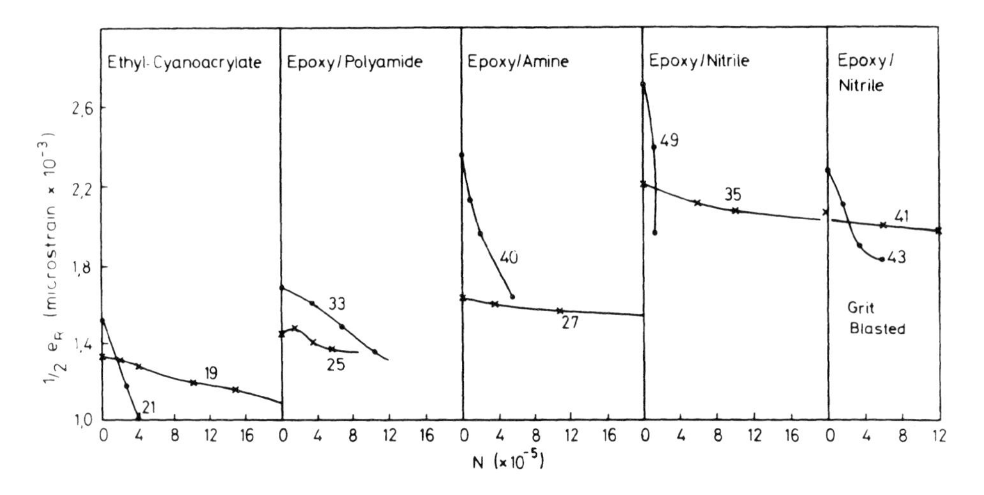

Figure 6.3. Plots of $\frac{1}{2}\, e_R$ versus number of cycles (N) for various adhesives, as indicated, used to bond a 0.15 mm thick b/ep reinforcement. The numbers in each plot indicate the nominal peak shear stress in the adhesive in MPa.

N with the value of e_s at the beginning of the test. Figure 6.5 plots effective length versus N, for the AF126 adhesive, using this approach.

The main practical conclusion from this study was that adhesive AF126 and, to a lesser extent, the epoxy/amine adhesive had useful properties for crack-patching up to about 80°C. However, since AF126 was also fairly insensitive to surface-treatment (see right-hand plot in Figure 6.3), it was chosen for most of our practical repairs despite its disadvantage of requiring to be cured at elevated temperature and its limited shelf life even under refrigeration. More recent work showed that film adhesives formulated for elevated temperature application, such as AF130 (by 3M Co) or FM300 (by Cyanamid) are suitable for some patching applications to well above 100°C.

6.2.2 Evaluation of minimum patch application processes and procedures

General considerations

Ideally, for ease of field application, the adhesives used for crack-patching should cure at reasonably low temperatures (above ambient) and pressures and should provide a high level of bond durability following a simple surface-treatment of the bonding surface.

The use of an adhesive curing at more elevated temperature is best avoided for several reasons, including (i) the difficulty, under field conditions of achieving and controlling locally elevated temperatures in a large structure, (ii) the development of undesirable residual and thermal-stresses and (iii) the possibility of undesirable metallurgical changes. Unfortunately, while useful mechanical properties for patching can be achieved with some adhesives curing at temperatures close to ambient (notably epoxies and modified acrylics), the structural film adhesives curing at elevated temperature (such as the epoxy-nitriles) appear to provide by far the best properties for high stress applications over a wide temperature range – from over 100°C to below −50°C.

Pressures between 100 and 300 kPa are required to obtain reasonable flow during cure of the above structural film adhesives – in contrast to paste adhesives, where usually only light holding pressure is required. Mechanical methods, for example hydraulic ram or screw jack, are the most desirable means of applying pressure; however, in many cases (including where there is danger of distorting the parent structure) these cannot conveniently be applied. The alternative then is to use atmospheric pressure by means of vacuum bag procedures. With this approach a nominal pressure of 100 kPa (or greater, with force multipliers) can be applied to a surface of any orientation. A further advantage is that heat can be applied by radiation through a nylon or teflon film by heat lamps or, alternatively, by electrical wires embedded in a reusable silicon rubber bag.

The vacuum bag approach, however, suffers from several major drawbacks. Besides the hazard of the bag developing an air leak at a critical stage in the repair, the drawbacks are mostly associated with the low pressures that may be created in some regions under the bag and include the danger that (i) large voids in the cured adhesive will result from expansion of air entrapped in the adhesive interface and

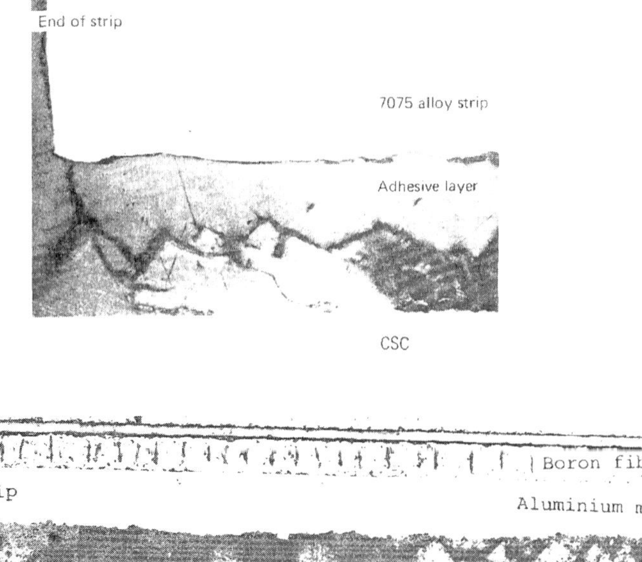

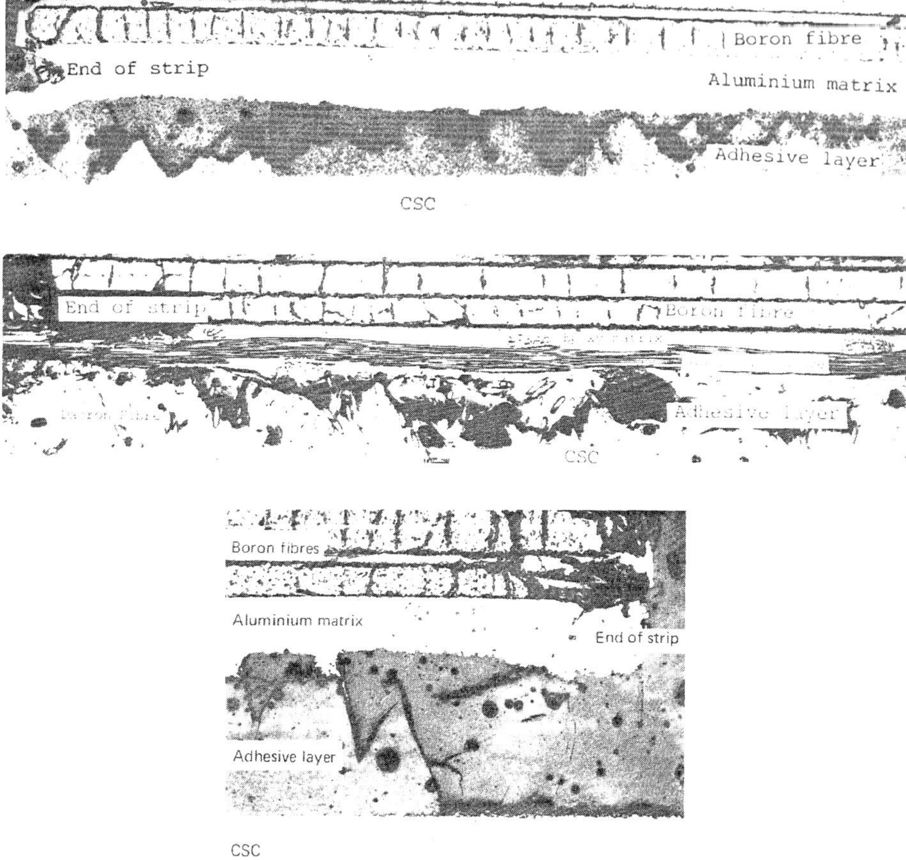

Figure 6.4. Micrographs of taper-sections through specimens with the following reinforcing strip/adhesive combinations, (a) 7075/cyanoacrylate, (b) boron aluminium/epoxy-polyamide, (c) boron epoxy/epoxy-nitrile and (d) boron aluminium/epoxy-amine. The adhesive thicknesses were of the order of 0.1 mm and the cyclic shear stress level were respectively ± 33 MPa, ± 36 MPa, ± 40 MPa and ± 40 MPa.

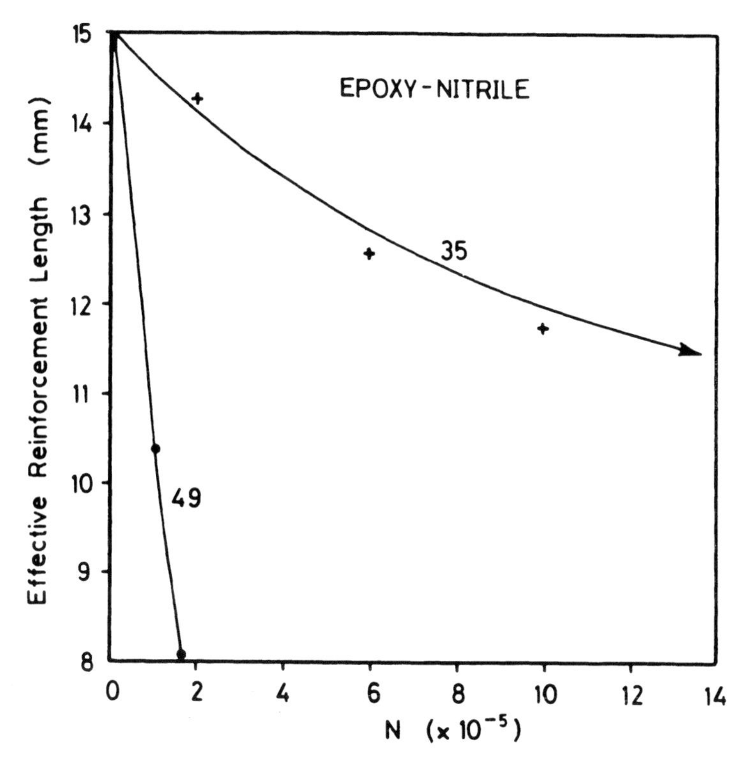

Figure 6.5. Effective reinforcement length versus cycles N for a b/ep reinforcement, initial length 15 mm, 0.15 mm thick, bonded with epoxy-nitrile adhesive AF126. The numbers on each plot refer to the nominal shear stress in the adhesive in MPa.

dissolved volatile materials – moisture, solvent, and (ii) air and fluid contaminants (such as hydraulic oils and water) may be drawn into the adhesive through the crack and through gaps – e.g. around fasteners, unless these regions are suitably sealed, for example with polysulphide rubber. Thus, vacuum bag procedures are generally unreliable for applying patches and are best avoided.

Finally, chemical surface-treatment of the metallic component is also best avoided, where feasible, for several reasons including (i) the difficulty of performing such treatments under field conditions, and (ii) the danger of residual or entrapped chemical agents encouraging stress-corrosion cracking from the pre-existing crack or of cracking or corrosion in adjacent areas – such as fastener holes or joints.

The following sections describe some experimental work aimed at developing minimum patch application procedures.

Minimum adhesive cure temperatures

Studies [9] were made to assess the minimum temperature at which satisfactory cure of the adhesive AF126 could be achieved; limited tests were also performed on adhesive FM73M (by Cyanamid). Both adhesives are nitrile-rubber modified epoxies, recommended cure is 1 hr at 120°C under a pressure of 0.3 MPa. FM73 is

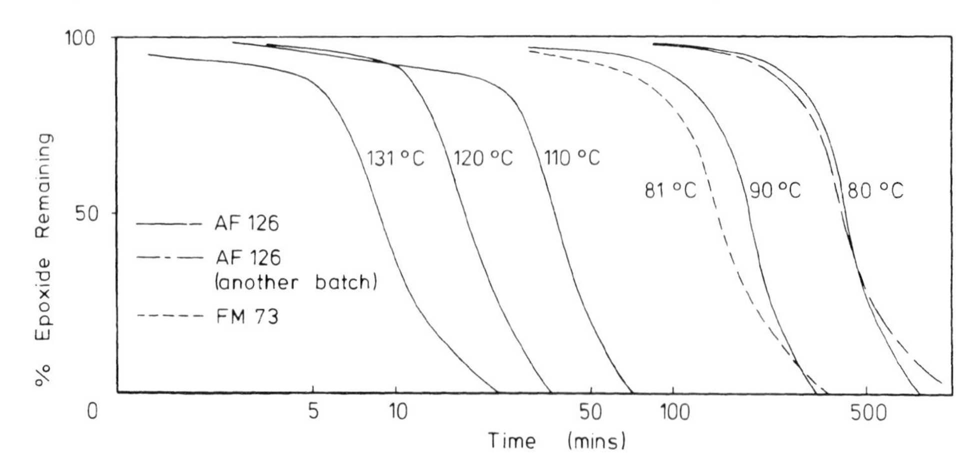

Figure 6.6. Rate of epoxide consumption from infrared studies at various temperatures for adhesives AF126 and FM73.

a state-of-the-art adhesive with improved storage life under refrigeration and increased working life at ambient temperature compared to AF126. In part, the improvement results from the increased resistance of FM73 to hydrolysis in the presence of atmospheric moisture.

The level of cure was assessed using infrared spectrography to follow the progress of the cure reaction. Spectra were run at intervals on a sample mounted in the spectrometer in an assembly maintained at the required temperature; the disappearance of the epoxide band (915 cm^{-1}) was taken as an indication of apparent cure. The results of this experiment, plotted in Figure 6.6 lead to the minimum effective cure times listed in Table 6.3. The rate of reaction of FM73 differs from AF126 because the two systems contain different curing agents.

Other tests employing differential thermal analysis, dynamic mechanical analysis and scanning calorimetry [9] showed similar nominal Tg values for the various cure conditions.

Finally, tension tests were performed on single overlap specimens made with these adhesives. The results for tests performed at 20 or 80°C, (Table 6.4), show

Table 6.3. Times for consumption of epoxide for adhesives AF 126 and FM 73.

Temperature, °C	Time for consumption of epoxide
AF 126	
80	> 12 h
90	5.5 h
110	70 min
120	35 min
131	23 min
FM 73	
81	6 h

Table 6.4. Nominal tensile shear strength of aluminium-aluminium single overlap joints prepared from AF 126 or FM 73 cured under various conditions.

Cure schedule	Shear strength, MPA	
	20°C	80°C
	AF 126	
1 h/121°C	28.4	21.3
8 h/80°C	24.6	14.0
15 h/80°C	27.4	21.3
	FM 73	
1 h/121°C	39.2	30.4
8 h/80°C	37.3	27.2
15 h/80°C	38.1	30.0

that for cure times consistent with Table 6.3, only a small penalty is incurred by use of the low cure temperature. Note the poor result at elevated temperature for the 8h/80°C cure of AF126 where 25–30% of the epoxide groups would have been unreacted.

Surface-treatments for the composite patch

Pre-cured composite patches require surface-treatment to ensure effective adhesive bonding. Fortunately, all that is usually required is a freshly produced epoxy surface free from surface contamination. The main source of contamination, on a recently moulded patch, will usually be release material transferred from the moulding surface. The moulding surface could be, for example, (i) a woven glass cloth coated with PTFE, (ii) a PTFE film material, (iii) a silicon rubber sheet, or (iv) a metal surface coated by spraying with either a PTFE dispersion or a silicone-containing solution. Surface-treatment consists, essentially, of removing this contamination by a mechanical procedure, this could involve (a) surface removal by a peel ply technique, (b) surface abrasion – for example with silicon carbide grit paper, or (c) surface erosion – for example by blasting with alumina grit.

Peel ply consists of a woven cloth, usually nylon, which is incorporated into the surface of the component. Prior to bonding, the ply is peeled from the composite, fracturing the epoxy polymer where it has extended through the weave, thus producing a clean active surface. There are, however, drawbacks to this procedure; (a) the resulting surface is usually very rough and tends to entrap air bubbles during bonding, and (b) there is a serious danger that small amounts of peel ply may be left on the surface and act as a release film – some sub-microscopic transfer of the ply material to the surface is unavoidable. Thus it is unwise to use the peel ply approach alone, abrasion treatments to effect surface removal and levelling should also be employed.

A comprehensive study on the influences of various types of release material and surface removal procedures on gr/ep bonded joints [10] showed that (i) PTFE coated glass cloth release produced the lowest surface contamination following moulding of gr/ep laminates, and (ii) grit-blasting with alumina particles almost completely removed any residual contamination. In most of our studies including the fatigue experiments referred to earlier, it was found that surface-treatment by alumina grit-blasting was sufficient to provide strong adhesive bonds to b/ep.

Minimum surface-treatments for aluminium alloy 2024-T3

Standard adhesive cure conditions

Our initial work with adhesive AF126 [8] indicated that alumina grit-blasting provided an adequate level of bond strength and durability for repairs not subjected to severe environmental conditions. Indeed, early practical repairs based on this approach have given over nine years of aircraft service without evidence of bond failure.

Our initial studies on improved surface-treatment procedures [11] were based on adhesive AF126 and AF130 cured at the (near standard) conditions of 120°C/1 hr/340 kPa and 177°C/2h/280 kPa respectively. Table 6.5 provides details of the surface-treatments evaluated on aluminium alloy 2024 T3 clad. Treatments involving the use of chromate solutions were avoided because of the danger of encouraging stress-corrosion cracking from the pre-existing crack. The PANTA anodising procedure was developed by Boeing Aircraft Co. [12] for field repairs to adhesively bonded aluminium. This procedure allows local anodisation, in a phosphoric acid solution, of an aluminium surface (at any inclination) on a large structure. To prevent run off, the acid is formed into a gel (with fumed silica) and supported by an open weave cloth which is sandwiched against the aluminium surface with a stainless steel wire screen cathode. The phosphoric acid tank anodise procedure was chosen to represent a standard factory surface-treatment, against which the others could be compared.

The potential durability of the adhesive bond provided with the candidate surface-treatments was assessed by means of the Boeing wedge test procedure,

Table 6.5. Details of the surface-treatments for aluminium alloys initially investigated to establish minimum procedures for acceptably durable adhesive bonding.

Treatment	Steps in process
Hand abrade (HA)	Vapour degrease, abrade with 'Scotchbrite A' until matt surface achieved, wipe clean with lint and grease free cotton cloths.
Grit blast (GB)	Vapour degrease; thoroughly grit-blast surface using 50 μm Al_2O_3 with dry N propellant; blow clean with dry N_2.
Phosphoric acid gel anodise (PANTA)*	Vapour degrease; abrade with 'Scotchbrite A' or grit blast with 50 μ Al_2O_3 and blow off debris; anodise at 6 volts for 10 minutes using phosphoric acid gel; water rinse; hot air dry.
Phosphoric acid anodise (TA)	Vapour degrease; alkaline clean; water rinse; deoxidise in FPL etch; water rinse; H_3PO_4 solution anodise 20–25 minutes at 10 volts; water rinse; hot air dry.

*Phosphoric acid non-tank anodise.

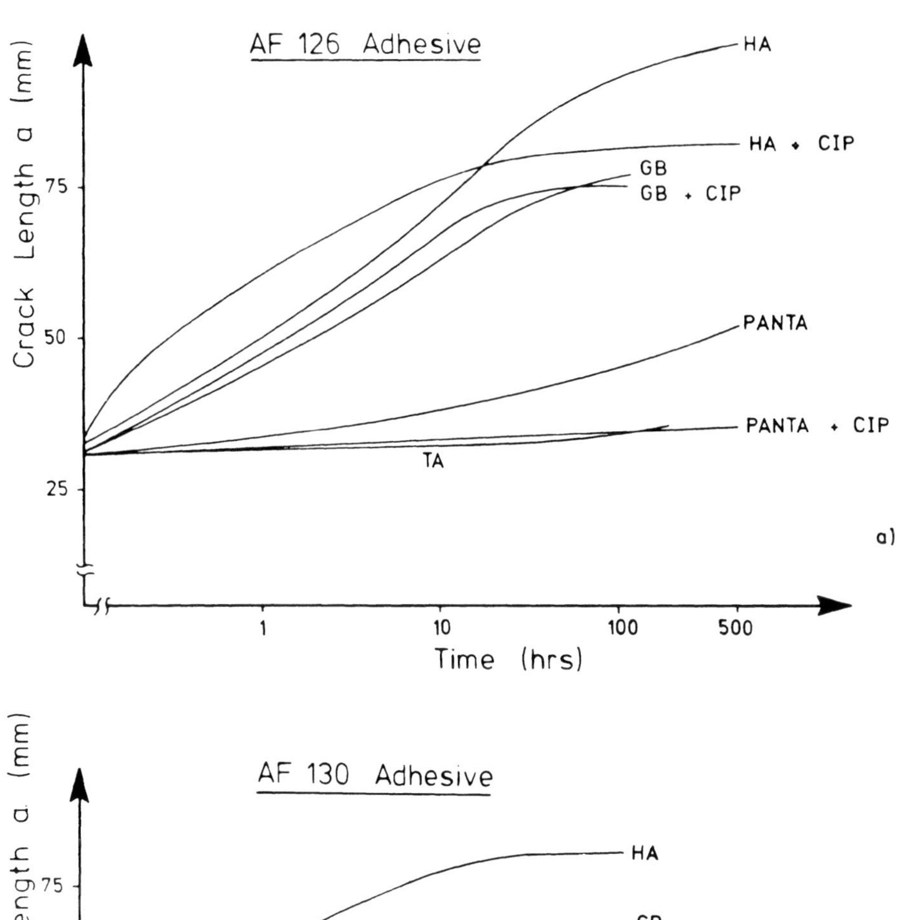

Figure 6.7. Plots of crack length versus time for wedge-test specimen made with (a) adhesive AF126 and (b) Adhesive AF130 following various surface-treatments, as indicated. Inset, schematic of wedge-test specimen.

illustrated in Figure 6.7(b) inset. In this test the crack growth in the bond zone, measured as a function of time, is used as an indicator of service durability. Aluminium alloy 2024 T3, clad (3.2 mm thick) was used for the adherends and the exposure conditions were nominally 50°C and 95% relative humidity.

The results for adhesive AF126 (Figure 6.7(a)) show that mechanical treatments give relatively poor resistance to crack-growth, the grit-blasting procedure being a little superior to the hand abrade – a surface-treatment procedure often employed in repair work. However, as expected, the PANTA procedure provides excellent resistance to crack-growth, particularly when the anodised surfaces are primed with a corrosion inhibiting epoxy primer (CIP, BR-127, by Cyanamid) prior to bonding. For wedge-test specimens prepared with adhesive AF130 (no primer) the results (Fig. 6.7b) indicate that this relatively brittle adhesive is significantly less sensitive to surface-treatment, since the GB treatment gives quite acceptable resistance to crack-growth and the PANTA treatment provides similar resistance to TA, the factory procedure.

Following the success of the phosphoric acid treatment, simple experiments were performed to establish whether entrapment of acid in the crack could promote crack growth by stress-corrosion. The experiments involved wedge cracking 7076 T6 aluminium alloy specimens and immersing them in the acid gel. No evidence was found to indicate that the gel encouraged cracking in this stress-corrosion prone material.

On the basis of the above studies the phosphoric acid gel anodising procedure was adopted for all critical applications of crack-patching to the present time. However, more recent work has shown that further simplification of the surface-treatment and curing conditions is possible.

Minimum surface-treatment and adhesive cure conditions

Studies were continued with the aim of simplifying further the bonding conditions for patch application. Work was aimed at evaluating (a) the influence of reduced cure temperature on bond durability, and (b) the use of silane primers to improve bond durability and to simplify surface treatment procedures.

Silanes are extensively employed as coupling agents to improve the durability of bonds between polymers such as adhesives and inorganics such as glass or metals [13]. Briefly, the silane molecule on hydration in water can be represented by the following simplified formula

$$R{-}{-}{-}{-}Si(OH)_3$$

The $Si(OH)_3$ group is thought to react with the oxide film at the surface of the inorganic to form a chemical (oxane) bond leaving the organo functional group represented by R free to react with the adhesive during its subsequent cure. Thus for our purposes R must be a group which is chemically compatible with the particular adhesive employed. The oxane bond can be represented by the formula

$$M{-}O{-}Si$$

where M = Al, Fe, Ti etc. Although hydrolysis of the oxane bond can occur in the presence of water, the reaction is reversible such that a degree of adhesion is always maintained even in the presence of water.

The silane used to promote bonding of epoxy adhesives to aluminium is usually γ-glycidoxypropyltrimethoxysilane (γ-GPS) in which R is an epoxy grouping which is compatible with epoxy adhesives. Reference [14], for example, showed that application of the silane primer to aluminium alloy surfaces prepared by a cold-etching procedure (based on chromic acid), markedly improved bond durability to a level almost equal to that obtained by phosphoric acid anodising.

In our study [15] Union Carbide silane A-187 (γ-GPS) was evaluated on clad 2024T3, whose surface was prepared either by phosphoric acid gel anodising or by grit-blasting (according to the conditions listed in Table 6.5). The silane treatment was as follows

(i) Scotch-brite abrade to remove scratches.
(ii) MEK wipe.
(iii) Grit-blast (50 μm alumina).
(iv) Phosphoric acid gel anodise (PANTA), in some cases only.
(v) Immerse in 1% by volume silane solution in distilled water.
(vi) Air dry and apply adhesive.

Adhesives AF126 or FM73M were used to form the wedge-test specimens and were cured at 80°C, 100°C or at the standard 120°C, under either mechanical pressure or vacuum bag conditions.

The results of wedge-test evaluation on these specimens are provided in Table 6.6 and in Figure 6.8. Low crack growth in the adhesive during the wedge test is the main indication of high bond durability. However, due to the configuration of the specimen (double cantilever) the driving force or energy release rate G for crack growth is a rapidly reducing function of crack size a. Thus the initial (maximum) value of energy release rate G_0 varies according to the initial crack size a_0; see Figure 6.7(b) inset.

A more rational basis of comparing the results from wedge test specimens is to compare the initial energy release rate G_0 normalised to a standard a_0 of 25 mm with the final value after 48 hours exposure, G_{48}. Unfortunately, as discussed in [16] accurate G values in the wedge-test are only obtained when G is relatively low ($G \approx 1$–$2\,\mathrm{kJ/m^2}$), otherwise significant errors arise from yielding of the metal adherends. Nevertheless, even though the G values obtained were relatively high in these tests, ($G_0 \approx 3$–$7\,\mathrm{kJ/m^2}$, AF126 and $G_0 \approx 2$–$4\,\mathrm{kJ/m^2}$, FM73M), the ratio G_{48}/G_0 is used in Table 6.6 and in Figure 6.8 as a useful means of comparing surface-treatments.

The results for AF126 show that, for surface-treatment by the PANTA procedure, durability is good with the 100°C and 120°C cure but is significantly reduced for the 80°C cure. As noted previously, durability is very poor when grit-blasting alone is employed for the surface-treatment. Use of the silane primer, however, appears to be highly beneficial, particularly for specimens initially surface-treated by grit-blasting.

FM73M appears to provide a reduced level, compared to AF126, of durability with PANTA process, particularly at the lower cure temperatures. However, the improvement gained by using the silane primer with either surface-treatment is quite dramatic.

Table 6.6. Data from wedge-test experiments for 2024 T3 clad specimens prepared with adhesives AF 126 or FM 73 made under the conditions indicated.

Adhesive	Cure condition	Surface-treatment	Pressure method	Initial crack size mm	Growth a (mm) in t hours				G_{48}/G_0
					2	4	24	48	
AF 126	80°C/8h	An	M	30	2.8	3.5	7.0	9.4	0.18
	80°C/8 h	An	V	35	2.3	3.7	7.0	8.8	0.13
	80°C/8 h	An + S	M	36	1.7	2.3	2.9	3.4	0.19
	100°C/4 h	An	M	30	0.9	1.2	2.2	2.4	0.42
	100°C/4 h	An	V	35	2.0	2.6	4.4	4.6	0.18
	100°C/4 h	An + S	M	30	1.1	1.6	2.9	3.8	0.35
	120°C/1 h	An	M	29	0.9	1.3	2.8	3.4	0.42
	120°C/1 h	An	V	35	1.0	2.2	2.8	3.4	0.22
	120°C/1 h	An + S	M	27	1.4	1.5	2.1	2.6	0.57
	120°C/1 h	GB	M	28	15	19	23	43	0.02
	120°C/1 h	GB + S	M	27	20	2.9	7.0	8.7	0.28
FM 73	80°C/8 h	An	M	37	8.7	19	27	33	0.02
	80°C/8 h	An	V	39	12.5	18	29	32	0.02
	80°C/8 h	An + S	M	38	1.1	1.4	2.3	2.9	0.17
	80°C/8 h	GB + S	M	40	10	1.4	2.6	3.3	0.14
	100°C/4 h	An	M	36	2.1	3.8	12.0	18.0	0.06
	100°C/4 h	An	V	36	4.0	6.8	17.5	19.4	0.05
	100°C/4 h	An + S	M	36	0.2	0.5	2.0	2.4	0.22
	100°C/4 h	GB + S	M	37	0.7	0.9	1.4	1.6	0.21
	120°C/1 h	An	M	36	1.7	1.8	6.8	9.6	0.11
	120°C/1 h	An	V	34	1.7	2.2	11.0	15.5	0.08
	120°C/1 h	An + S	M	34	1.2	1.5	2.4	2.6	0.26
	120°C/1 h	GB	M	28	15.2	18.7	33.2	42.6	0.02
	120°C/1 h	GB + S	M	35	0.4	0.8		1.4	0.27

Note: Five specimens tested for each condition.
Notation: An PANTA, GB grit-blast, S silane prime, M mechanical pressurisation, V vacuum bag pressurisation.

From the above results, it is concluded that adhesives FM73, and possibly AF126, will provide excellent durability with an 80°C cure following surface-treatment of the metallic adherends by grit-blasting and a silane primer.

6.3 Thermal and residual-stress problems

Despite the numerous advantages of b/ep and gr/ep for repair patches, the potential disadvantages resulting from residual and thermal stresses must be considered. Residual stresses (tension in the metal, compression in the patch) result from the difference in coefficients of thermal expansion between the metal and composite patch when adhesives curing at above ambient temperatures are employed. Ther-

Figure 6.8. Histogram representation of the data presented in Table 6.6 for mechanical pressurisation for adhesives (a) AF126, and (b) FM73M. The percent figures indicate the approximate degree of cohesive failure (failure in the adhesive layer) for each treatment.

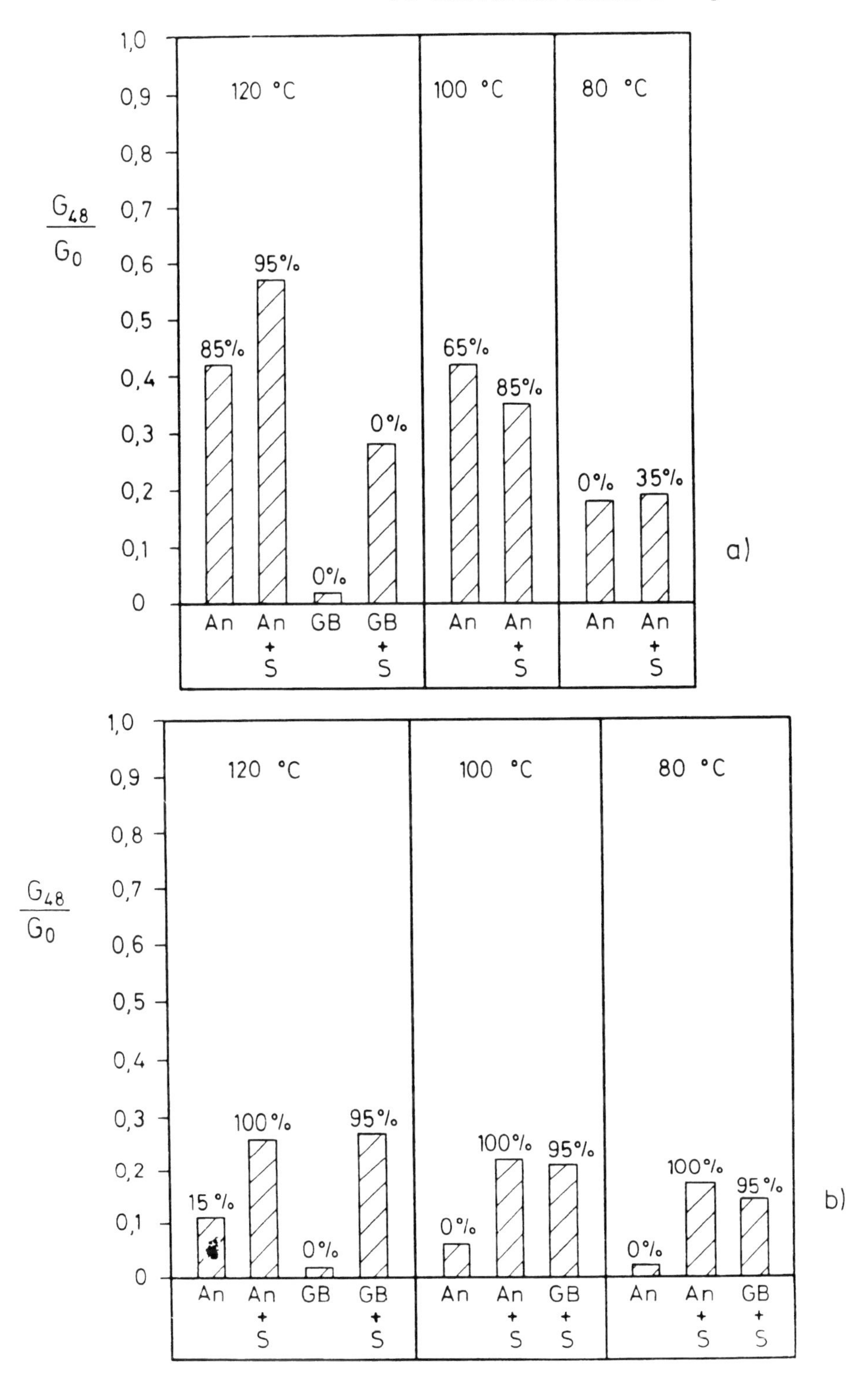
1,0
0,9
0,8
0,7
0,6
0,5
0,4
0,3
0,2
0,1
0
$\frac{G_{48}}{G_0}$
120 °C
100 °C
80 °C
85%
95%
0%
0%
65%
85%
0%
35%
An
An + S
GB
GB + S
a)
15 %
100 %
95%
100%
95%
100%
95 %
0%
b)

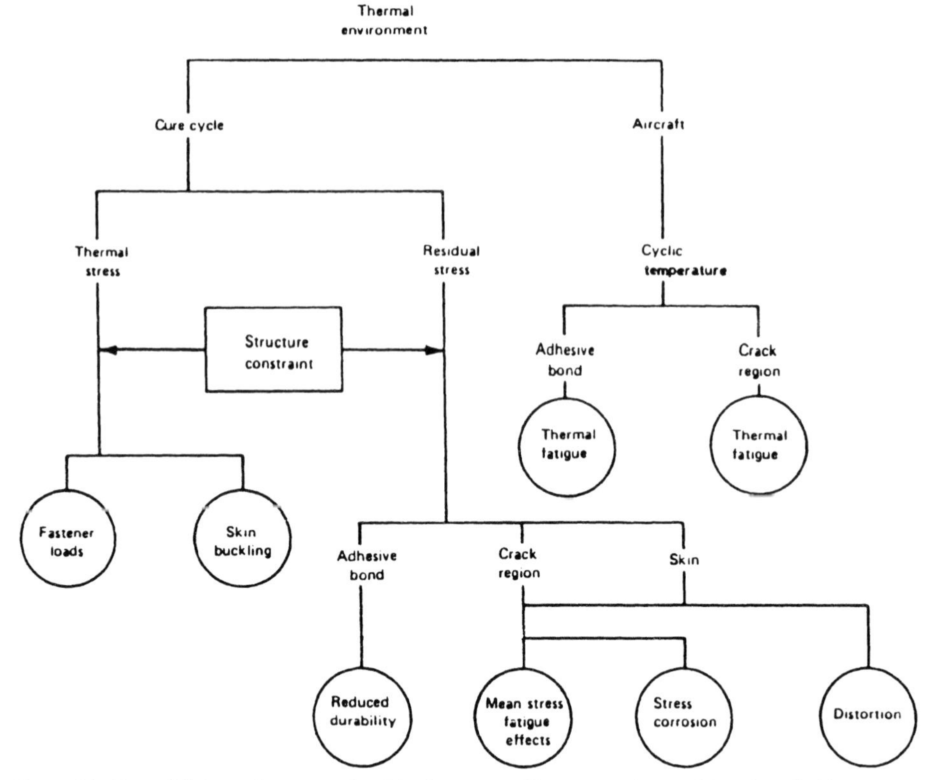

Figure 6.9. Potential thermal-stress and residual-stress problem areas when making aircraft skin repairs with advanced fibre composite patches.

mal stresses result when a local area of a large component is heated during application of the patch, because expansion of the heated area is restrained by the surrounding cold structure. Figure 6.9 categorizes some of the difficulties which may arise as a result of these stresses.

Experiments were conducted [17, 18] to study the following aspects which were considered to be of particular concern:

1. The magnitude of the stress-intensity factor K_T caused by patching.
2. Influence of restraint by the surrounding structure on thermal and residual stress.
3. Influence of residual stress, following patching, on the fatigue properties of the metal structure.
4. Influence of thermal cycling on the adhesive system.

Only aspects (i) and (ii) will be discussed in detail here; only a few comments are made here on our findings with respect to aspects (iii) and (iv).

To study the influence of the residual stress, fatigue tests were performed [17] on (uncracked) aluminium alloy specimens reinforced with either b/ep or gr/ep bon-

Figure 6.10. Plot of (a) measured stress at various points along the aluminium core of the reinforced specimen, (b) predicted stress $\sigma_T(y)$ obtained from equations 6.1, for two values of shear modulus, and (c) schematic of specimen indicating the notation used for thicknesses and lengths.

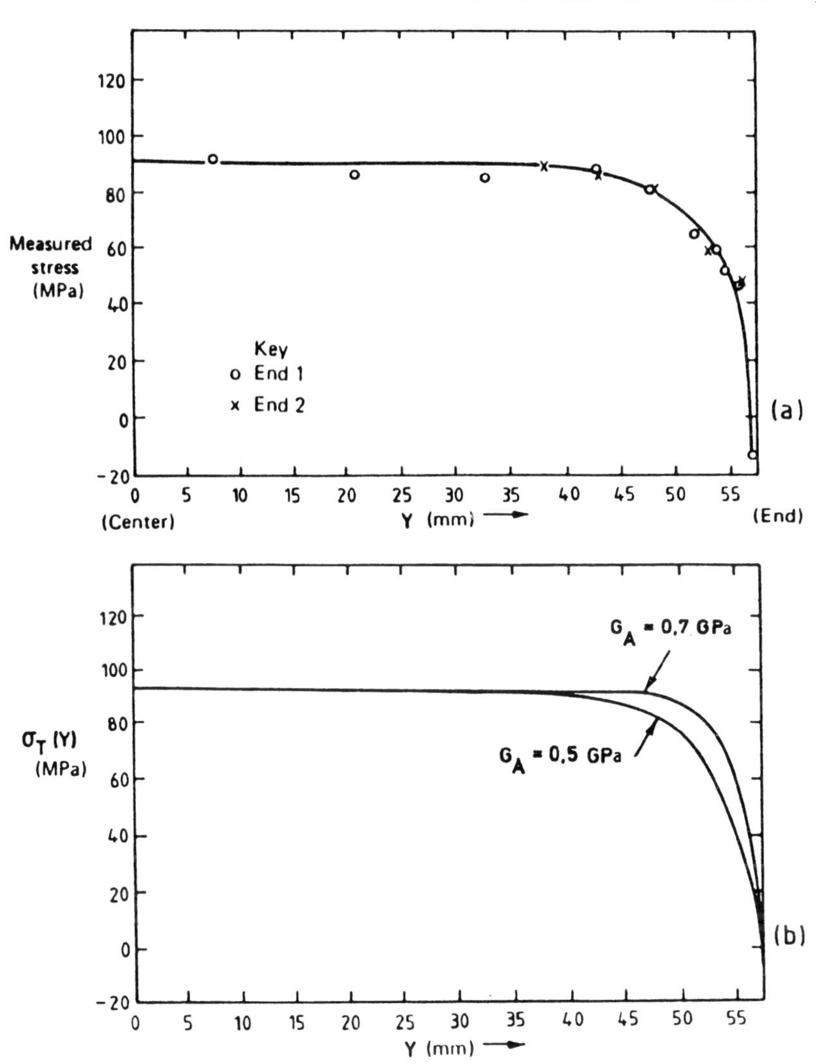
120
100
80
Measured stress (MPa)
60
40
20
0
-20
Key
o End 1
x End 2
(a)
0
5
10
15
20
25
30
35
40
45
50
55
(Center)
Y (mm)
(End)
G_A = 0,7 GPa
G_A = 0,5 GPa
σ_T (Y) (MPa)
(b)
Y (mm)

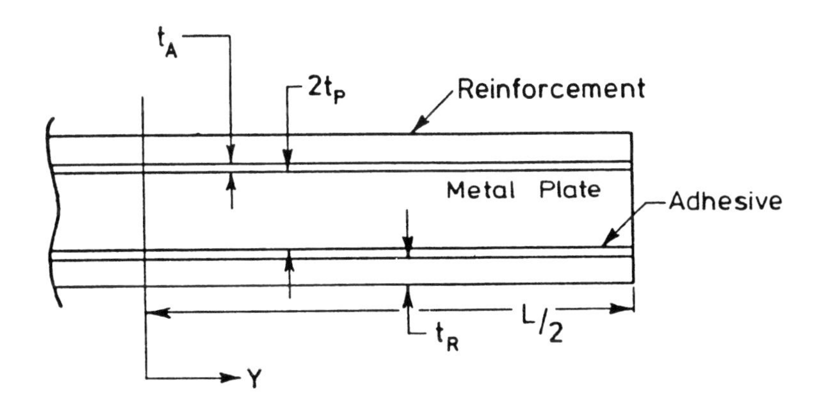
t_A
$2t_P$
Reinforcement
Metal Plate
Adhesive
t_R
L/2
Y

ded with adhesive AF126, cured at 120°C. It was found that the residual stress produced in the metal by patching had a small, but significant, adverse effect on the fatigue life of the metal when compared with the unreinforced metal at a similar strain level. However, the reduction in life due to the (residual) mean stress, was masked by the beneficial influences of (a) surface-treatment of the metal by grit-blasting, and (b) isolation of the metal from the environment by the patch system. It was also noted that, as expected, the residual compressive strain in the reinforcement effectively improved its tensile strain to failure. The influence of residual-stress on cracked specimens is discussed later.

To study the influence of thermal cycling on the adhesive system, tests were performed [18] in which aluminium alloy specimens, reinforced with the b/ep-AF126 system were cycled (for over 1000 cycles) between +120°C (the nominal stress-free temperature) and temperatures down to −50°C, representative of a severe condition which could be experienced by a metal aircraft structure at high altitude. Under these conditions, no cracking by thermal fatigue of the adhesive was observed. Cracking in the adhesive was only observed when the unrealistically low minimum temperature of −120°C was imposed during the thermal cycle.

6.3.1 Estimation of the residual stress-intensity, K_T

Residual stresses in a patched structure will give rise to a residual stress-intensity K_T. To assess this, a study involving the use of X-ray back reflection procedure was undertaken [18], in which aluminium alloy (7075-T6) sheet specimens (1.5 mm thick) were reinforced with gr/ep sheets (0.49 mm thick) covering both surfaces, Figure 6.10(c). Graphite/epoxy composite was chosen rather than b/ep to allow the X-rays to penetrate through to the metal surface. The composite reinforcements were bonded to the metal sheet with AF126, cured at 126°C. Some of the aluminium alloy sheets had simulated center cracks (either 5, 15 or 30 mm long) oriented normally to the fibres. The simulated cracks were produced by spark machining and had tip radii of about 0.2 mm.

The theoretical value for the peak residual stress σ_T in the uncracked metallic component [17] is given by

$$\sigma_T = t_R E_R E_P \Delta T(\alpha_P - \alpha_R)/(t_P E_P + t_R E_R) \qquad (6.1a)$$

and the corresponding stress distribution $\sigma_T(y)$ in the fibre direction is given by

$$\sigma_T(y) = \sigma_T(1 - \cosh \beta y/\cosh \beta L/2) \qquad (6.1b)$$

where the lengths y and L and the thicknesses t_P and t_R are as indicated in Figure 6.10(c), ΔT is the temperature of cure minus ambient temperature, α_P and α_R are respectively the expansion coefficient of the metal plate and reinforcement, E_P and E_R are respectively the Young's modulus for the plate and reinforcement. The elastic strain exponent β is given by

$$\beta = \{(G_A/t_A)(1/t_P E_P + 1/t_R E_R)\}^{1/2} \qquad (6.1c)$$

Figure 6.11. Plot of (a) stress, σ_y, versus distance from the crack tip, r measured along the axis of the crack as indicated, (b) apparent stress-intensity, K, estimated from the above plot, versus distance from the crack tip, r.

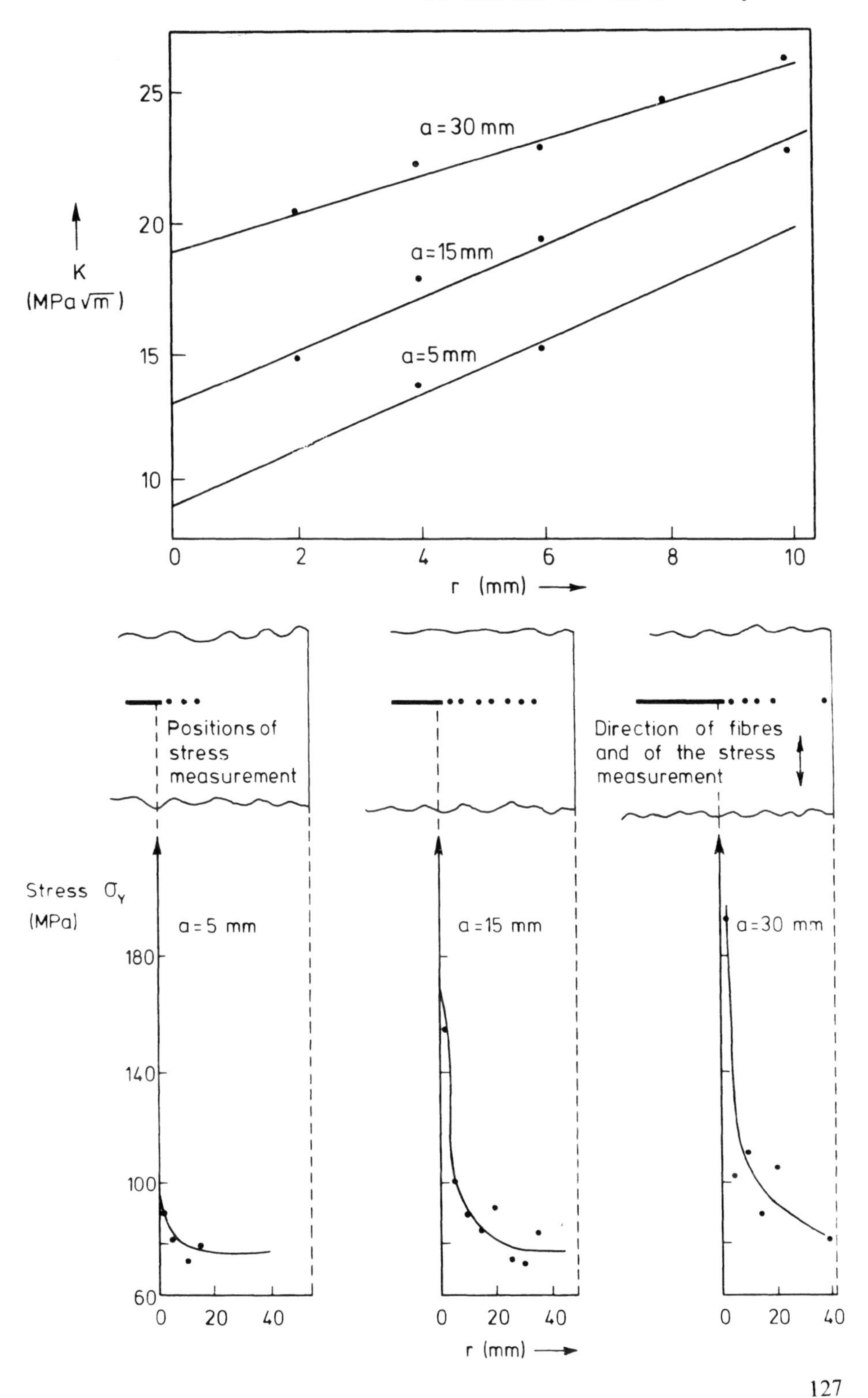

25
20
15
10
K
(MPa √m)
a = 30 mm
a = 15 mm
a = 5 mm
0
2
4
6
8
10
r (mm)
Positions of stress measurement
Direction of fibres and of the stress measurement
Stress σ_Y
(MPa)
a = 5 mm
a = 15 mm
a = 30 mm
180
140
100
60
0
20
40
r (mm)

where G_A is the effective shear modulus of the adhesive. The distance β^{-1} is the characteristic load transfer length (β is actually the exponent of the elastic shear strain distribution) and is usually about one third the distance from the ends of the reinforcement to where the strains in the plate and reinforcement become approximately constant or the shear stress in the adhesive layer becomes approximately zero.

In the thermal-fatigue work mentioned earlier the peak thermally induced shear stress in the adhesive layer τ_T (assuming elastic behaviour) is approximately given by

$$\tau_T = \sigma_T t_P \beta. \tag{6.2}$$

The X-ray measurements in plain (uncracked) specimens, Figure 6.10(a), showed that the observed peak residual-stress, was in excellent agreement with the theoretical value, Figure 6.10(b), and also that the observed residual-stress distribution was in fair agreement with $\sigma_T(y)$ if a shear modulus G_A of about 0.5 MPa was assumed for the adhesive.

Results for the artificially cracked aluminium alloy sheets are given in Figure 6.11(a) as a plot of measured stress, σ_y, in the fibre direction, against distance r, in direction x from the tip of the crack, for each of the crack sizes studied. The apparent stress-intensity K produced by the residual-stress was estimated from the data in Figure 6.11(a) by using the following relationship for a blunt crack [19]

$$\sigma_y = K/(2\pi r)^{1/2}(1 + \varrho/r)/(1 + \varrho/2r)^{3/2} \tag{6.3}$$

where ϱ is the radius of curvature of the crack tip; in this case, 0.2 mm. The apparent value of K_T was obtained from these results by firstly plotting K versus r, Figure 6.11(b), and then, by extrapolation, finding K_T as $r \to 0$.

Using the above approach it was found that the stress-intensities measured experimentally could be expressed as

$$K_T \simeq \sigma_T(\pi a)^{1/2} \tag{6.4}$$

where $2a$ is the crack length and σ_T is given by equation (6.1).

Whilst, on first consideration, this relationship appears reasonable for K_T it was unexpected on the basis of theoretical consideration of the problem. It was originally conjectured [17], based on strain energy considerations, that the stress intensity (due to the residual stress) should be constant with respect to crack length. Rose [20] reached a similar conclusion following a more rigorous analysis of the problem in which a characteristic crack size was identified at which K_T approaches a constant value K_α. However, the characteristic crack size for the configuration described here is much less than 30 mm. Thus it appears that while further studies are required to resolve this aspect, it is at least conservative to assume that, for fairly small cracks, K_T is as given approximately by equation (6.4).

6.3.2 *Influence of restraint by the surrounding structure*

Restraint of the heated region by the surrounding cool structure during patch application gives rise to thermal stresses which may result in some of the problems noted in Figure 6.9. However, restraint will also be highly beneficial in lowering the level of residual stress by reducing the effective expansion coefficient of the metal in the hot zone. The level of restraint depends largely on the stiffness of the surrounding structure. In typical aircraft structure the level of restraint is generally relatively high due to the size of the structure and the rigid fastening between its component parts.

The above aspect was studied experimentally [17] and analytically [20, 21]; Chapter 5 gives more detailed analytical consideration. In reference [20] a simple one-dimensional mathematical model of the patching region was considered, Figure 6.12(a), in which (i) the edges of the component at $\pm L_P$ are firmly clamped and act as heat sinks, (ii) the temperature is held constant at T over the patching region $2L_R$, and (iii) the temperature decays by ΔT linearly from the end of the heated zone to the heat sink – as expected for steady state conditions. It is shown that the effective coefficient of expansion α'_P is then given by

$$\alpha'_P = (\alpha_P/2)(1 - L_R/L_P). \tag{6.5}$$

This relationship shows that α'_P has a minimum value of zero when $L_R/L_P \to 1$, as expected, and has a maximum value of $\alpha_P/2$ when $L_R/L_P \to 0$.

Taking the model for the patching region to be a clamped circular plate, Figure 6.12(b) and assuming otherwise similar conditions to the one-dimensional model it is shown in reference [21] that the effective expansion coefficient α'_P is given by

$$\alpha'_P = \alpha_P(1 + \nu_P)[\tfrac{1}{2} - L_R^2/2L_P^2 + [(L_R^2/2L_P^2) \ln L_R/L_P + (1 - L_R^2/L_P^2)/4 \ln L_R/L_P]. \tag{6.6a}$$

This relationship shows again that α'_P has a minimum value of zero when $L_R/L_P \to 0$ but has a minimum of $\alpha'_P = (\alpha_P/2)(1 + \nu_P)$ when $L_R/L_P \to 0$ which is about 30% larger than the value predicted by the simple one-dimensional model.

It was also shown in reference [21] that if the circular plate had free edges, then:

$$\alpha'_P = \alpha_P\{(1 + \nu_P)2 + (1 - \nu_P)[(L_R^2/L_P^2 - 1)/4 \ln L_R/L_P)]\} \tag{6.6b}$$

In this situation $\alpha'_P = (\alpha_P/2)(1 + \nu_P)$ when $L_R/L_P \to 0$, as for the fixed edge condition and $\alpha'_P = \alpha_P$ as $L_R/L_P \to 1$, as expected.

Experiments were performed [17] in which a sheet of aluminium alloy 2024T3 (one meter square and 2.5 mm thick) was reinforced on one side with a b/ep patch (152 mm long × 38 mm wide × 0.76 mm thick), bonded to the metal sheet with adhesive AF126 cured at 120°C. The size of the heated zone $2L_R$ was approximately 300 mm and $2L_P$ was 1000 mm. During bonding of the patch, the metal sheet was restrained by bolting it at its ends to a stiff steel frame. For comparison, another sheet of aluminium alloy of similar thickness to the above panel, but of the same dimensions as the patch, was entirely covered by a patch on one side. After patching, and cooling to ambient temperature the curvatures of the patched regions were measured, in the case of the large panel, after cutting the patched region out of the panel. These situations are represented as I and II in Figure 6.12(c). The effective coefficients of expansion of the metal sheet α'_P were found from the curvatures using bimetallic strip theory, assuming that the coefficient of

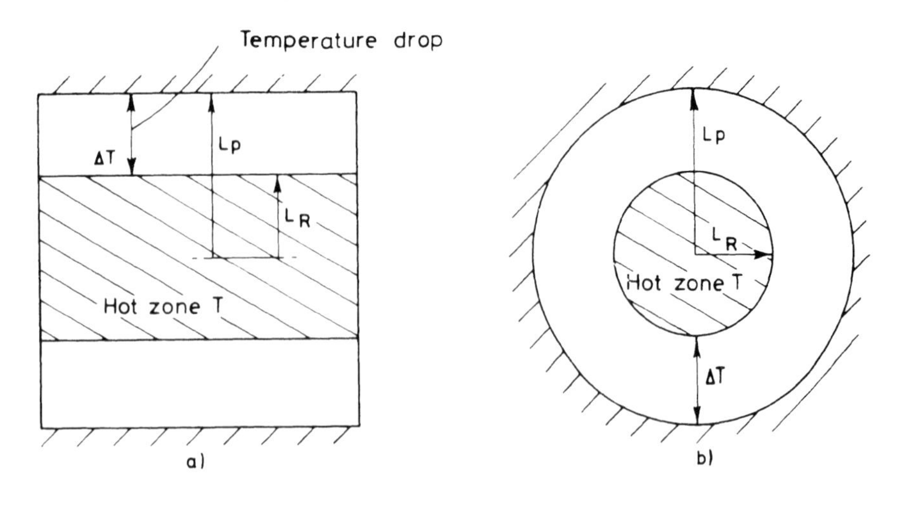

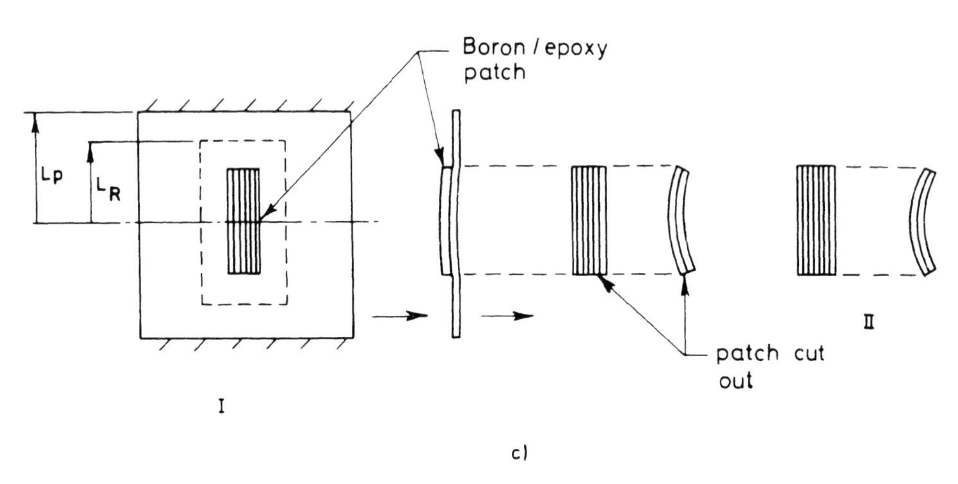

Figure 6.12. Schematic of (a) one dimensional model for estimating the influence of restraint to expansion during heating of a metal plate, (b) two-dimensional model, and (c) illustration of some simple experiments undertaken to measure residual-stress.

expansion of the patch in the fibre direction, α_R was as given in Table 6.1. The results obtained were:

(i) For the restrained situation $\alpha'_P = 14 \times 10^{-6}\,^{\circ}C^{-1}$

and (ii) For the unrestrained situation $\alpha'_P = 24 \times 10^{-6}\,^{\circ}C^{-1}$

The result for the unrestrained situation is in agreement with the expected value, given in Table 6.1. The predicted values for α'_P in the restrained case are respectively $\alpha'_P = 8.4 \times 10^{-6}\,^{\circ}C^{-1}$ using equation (6.5) and $\alpha'_P = 9.5 \times 10^{-6}\,^{\circ}C^{-1}$ using equation (6.6a), which differ significantly from the experimentally determined value.

The above discrepancies between theory and experiment could arise from several sources, including inadequate modelling of the experimental situation by the simple analytical models. However, probably the two major sources of discrepancy were (a) the restraint provided by the bolts was substantially less than perfect, and (b) the temperature distribution was not linear as assumed. In reference [17], using a one-dimensional model similar to Figure 6.12(a) with an empirical correction for the observed temperature distribution in the experiments, α'_P was found to be $11 \times 10^{-6}\,^{\circ}C^{-1}$ which is much closer to the experimentally determined value.

The apparent residual-stress in the restrained plate is obtained from equation (6.1) substituting α'_P (experimentally determined value) for α'_P giving $\sigma_T = 29\,MPa$. However, the value for σ'_P when the plate remains under constraint when cool (as it would do in the actual structure) is slightly lower than this, allowance for this effect [17] gives $\sigma_T = 24\,MPa$. These values should be compared with the unrestrained value of $\sigma_T = 60\,MPa$.

From the above it follows that if the repaired plate had been center-cracked, using equation (6.4), $K_T = 24\,MPa\,(\pi a)^{1/2}$ with the restraint compared to $K_T = 60\,MPa\,(\pi a)^{1/2}$ without restraint. Alternatively, the theoretical estimate using the clamped circular plate model, which is probably quite realistic for a well fastened aircraft structure, gives $K_T = 13\,MPa\,(\pi a)^{1/2}$.

6.3.3 Some approaches for minimising residual stress

Several approaches can be employed to minimise residual stress following patch bonding at elevated temperature including the following:

(i) The area heated should be minimised to maximise the constraint offered by the surrounding structure.

(ii) The adhesive (and if cocuring is employed, the patch) should be (a) cured at the lowest possible temperature – for example, adhesive FM73 can be cured at temperatures as low as 80°C, see section 6.2.2 or (b) pre-cured at the lowest reasonable temperature and then post-cured at a higher temperature.

(iii) If feasible, the patched structure should be pre-stressed in compression during patch application to a level which will partially or completely nullify the residual tensile stress on cooling.

It may even be possible to arrange for a net compression stress in the patched component. This approach will be applicable only on a structure, such as a wing, which is amenable to stressing e.g. by external jacks.

6.4 Design correlations and materials allowables

The viability and effectiveness of a patch repair can be assessed from

(i) The stress intensity factor range in the repaired component.

(ii) The shear strain range in the adhesive layer.

(iii) The tensile strain range in the reinforcing patch.

The stress intensity factor provides an indication of the likely rate of crack propagation following repair, whilst the other two parameters indicate the likely durability of the repair.

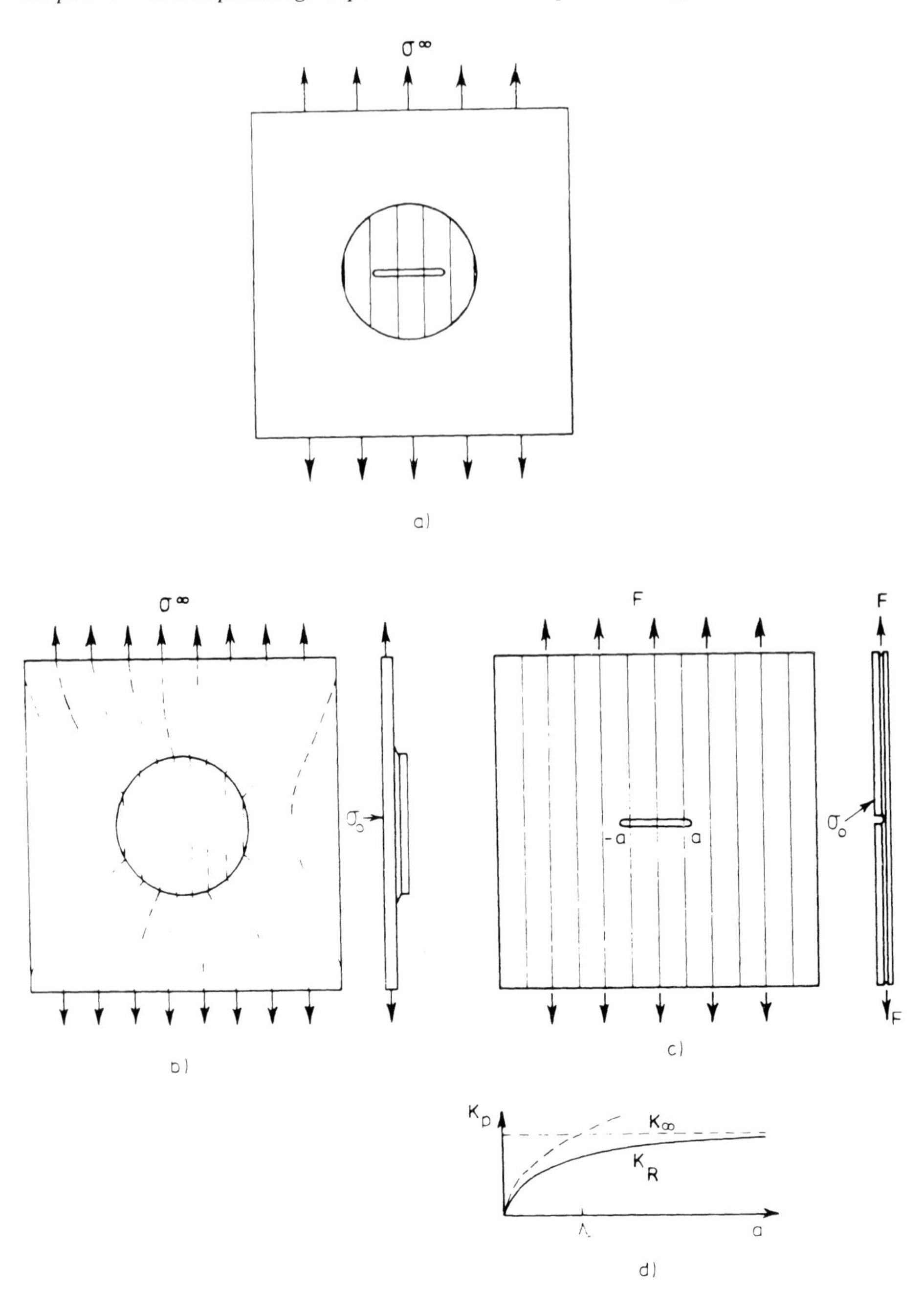

Figure 6.13. Schematic illustration of the analytical approach to crack-patching.

Two complementary design approaches have been developed in Australia. One is a finite-element procedure (Chapter 4) which has proved to be highly valuable in the design of several of our practical repairs. The other is an analytical procedure (Chapter 5) and is valuable in (a) providing a rapid feasibility estimate for a repair, and (b) indicating clearly the significance of the materials properties and geometrical parameters associated with the repair.

It is the purpose here to describe experimental work which provided (i) correlation with the analytical design predictions, and (ii) relevant materials properties and allowables. A proposed design procedure is described in section 6.5, based on some of these observations.

In the analytical procedure (described in detail in Chapter 5), predictions for the various parameters in the patching problems can be obtained from a two stage analysis as illustrated in Figure 6.13 for a center-cracked panel.

Briefly, the first stage of the analysis, assumed that the patch thickness t_R, was bonded to one side of an uncracked plate, thickness t_P, at the prospective location of the crack, Figure 6.13(b). The stress σ_0 under the patch was found by modelling the region as an inclusion in the plate which is subjected to a remote stress σ^∞. The second stage of the analysis assumed that the situation in the cracked region was as represented by Figure 6.13(c) where the reinforcement completely covers its surface and the applied force intensity (force per unit width) is F.

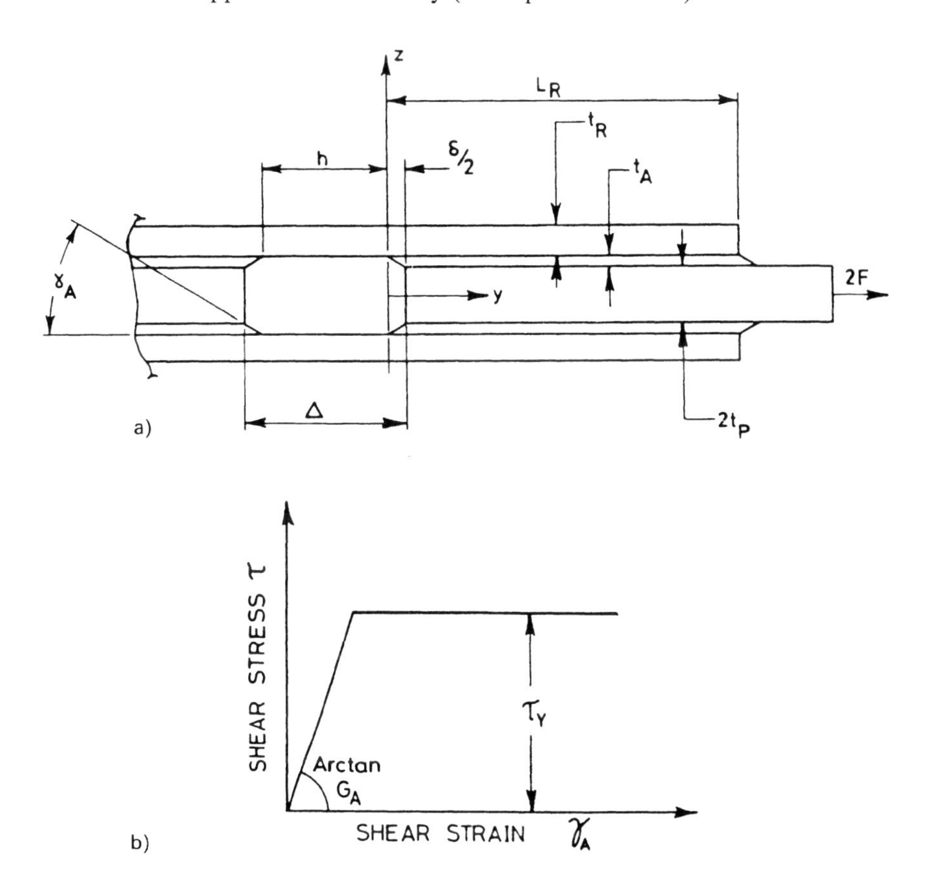

Figure 6.14. Schematic illustration of joint specimen indicating (a) notation used for thickness and displacements, and (b) elastic/perfect plastic stress strain behaviour assumed for the adhesive.

Using the above approach it is shown that conservative values for the key design parameters can be obtained by considering an overlap joint, of geometry as depicted in Figure 6.14(a), which is subjected to force intensity $2F$. As illustrated schematically in Figure 6.13(d), it is found that the stress-intensity following patching K_r tends towards a maximum value K_∞ which becomes independent of crack size beyond a characteristic crack length Λ.

An important point from the analysis is that the overlap joint is a much simpler system to study theoretically or experimentally than a patched panel and this leads to considerable simplification in assessing a repair scheme, especially when considering the effect of deviations from ideal behaviour, such as plastic or viscoelastic relaxation in the adhesive or environmental and fatigue degradation of the bond.

The parameter which is of greatest relevance to the theory is the relative displacement δ, because in the theory the displacement under the applied force is used to determine the upper bound for the crack extension force G_∞, which in turn is used to find K_∞. Also, δ is directly related to the shear strain γ_A in the adhesive layer, which is an important parameter with respect to the durability at the repair. The other important parameter influencing repair durability is the strain in the reinforcement e_R which is also obtained from consideration of the overlap joint.

6.4.1 Theoretical results

A. Overlap joint

This section summarizes the important analytical relationships for the overlap-joint based on the one-dimensional analysis given in chapter 5. In all these results it is assumed that the overlap length L is very long compared with the characteristic load transfer length, such that $\beta L \gg 1$

(i) *Relationships for δ and γ_A*

Examination of Figure 6.14(a) shows that

$$\gamma_A = \delta/2t_A. \tag{6.7}$$

For elastic behaviour of the adhesive $\tau < \tau_y$ Figure 6.14(b) it can be shown that that

$$\delta = 2F/(\beta E_R t_R) \tag{6.8}$$

where $S = (E_R t_R)/(E_p t_P)$.

However, if F exceeds F_y the force intensity at which the adhesive begins to yield, assuming elastic/perfectly plastic behaviour (Figure 6.14(b)) it can be shown that

$$\delta = \tau_y(t_A/G_A)[1 + (F/F_y)]^2, \quad F > F_y \tag{6.9a}$$

If τ_y is the effective shear yield stress in the adhesive, F_y is given by

$$F_y = (\tau_y/\beta)(1 + S) \tag{6.9b}$$

where $S = (E_R t_R)/(E_p t_p)$.

(ii) *Relationships for F and F' in terms of stresses*

If σ^∞ is the applied stress at the ends of the overlap joint then

$$F = t_P\sigma^{\infty} = t_P\sigma_0(1 + S) \tag{6.10a}$$

where σ_0 is the peak stress in the metallic adherend. If the metallic adherend is also subject to residual stress σ_T due to thermal effects, (as given by equation (6.1)) its effect on the joint can be allowed for by adding the internal force $\sigma_T(1 + S)$ to the external force. The equivalent force F' is then given by

$$F' = t_P[\sigma^{\infty} + \sigma_T(1 + S)] = t_P(1 + S)(\sigma_0 + \sigma_T). \tag{6.10b}$$

(iii) *Relationships for $\Delta\delta$ and $\Delta\gamma_A$*

The range of displacement and shear strains $\Delta\delta$ and $\Delta\gamma_A$ respectively are of interest with respect to cyclic loading. Specifically, $\Delta\delta$ is related to ΔK_{∞} the stress-intensity range suffered by the cracked component and $\Delta\gamma_A$ is expected to determine the fatigue damage suffered by the adhesive. For the patching situation when the minimum applied stress $\sigma^{\infty}_{\min}$ is zero and the residual stress σ_T is also zero then $\Delta\delta = \delta$ and $\Delta\gamma_A = \gamma_A$. However, where $\sigma^{\infty}_{\min}$ is zero but σ_T is not then

$$\Delta\delta = \delta - \delta_T \tag{6.11a}$$

$$\Delta\gamma_A = \gamma_A - \delta_T/2t_A \tag{6.11b}$$

where δ and γ_A are obtained from equation 6.8a if $F < F_y$ or equation 6.9a if $F > F_y$. The thermal displacement δ_T is obtained from equation 6.8 or 6.9, taking $F = \sigma_T t_P(1 + S)$.

(iv) *Relationships For Strains*

The peak strain in the reinforcement e_R occurs over the gap h, Figure 6.14 and is given by

$$e_R = F/(E_R t_R). \tag{6.12}$$

B. Patched plate

This section summarises the important analytical relationships derived in [22], for use in discussions here and in the design approach discussed in section 6.5.

(i) *Reduced Stress σ_0*

The reduced stress σ_0 at the location of the crack Figure 6.13 is obtained from the applied stress σ^{∞} through

$$\sigma_0 = \phi\sigma^{\infty} \tag{6.13a}$$

where ϕ is a factor depending on the shape and elastic properties of the patch. In the simplest case where the patch totally covers the panel, there is no inclusion effect (Figure 6.13(b)) and the situation is similar to the lap joint, then

$$\phi = (1 + S)^{-1}. \tag{6.13b}$$

For the case of main interest here the unidirectionally reinforced patch is modelled as an orthotropic inclusion of elliptical shape and allowance is made for biaxial stresses, then

$$\phi = (1 + S)^{-1}[1 + p + v_P S(\Sigma + q)] \tag{6.14a}$$

where ν_P is the Poisson's ratio for the metal sheet and Σ is the ratio $\sigma_x^\infty/\sigma^\infty$ and σ_x^∞ is the stress parallel to the crack direction. The values of p and q are obtained by solving the following simultaneous equation

$$\left[\frac{1 + 2(1 + S)\lambda R^{-1}}{\nu_p S - (1 + S)\lambda^{-2}} \middle| \frac{\nu_p S - (1 + S)\lambda^2}{1 + (1 + \nu_p^2)S + 2(1 + S)\lambda^{-1}R}\right]$$

$$\left[\frac{p}{q}\right] = S(1 - \nu_p\Sigma)\left[\frac{1}{-\nu_p}\right] \tag{6.14b}$$

where $\lambda = G_R/E_{Rx}$ and G_R and E_{Rx} are respectively the in-plane shear modulus and the transverse modulus for the patch and $R = L_R/W_R$, where L_R is the overlap length of the patch and W_R its half width. Finally, the effective force F (Figure 6.13(c)) is given by equation 6.10a

(ii) *Crack Extension Force* G_∞ *and Stress Intensity Factor* K_∞

The upper bound to G_∞ can be deduced by considering the work done in producing a displacement δ in an equivalent lap joint and is given by

$$G_\infty = F\delta/2(1 + S)t_P = \tfrac{1}{2}\sigma_0\delta \tag{6.15}$$

$$= (t_A/G_A)\beta t_P\sigma_0^2. \tag{6.15b}$$

The upper bound to K_∞ can be derived from G_∞ using the usual relationship for generalised plane stress

$$K_\infty = (E_P G_\infty)^{1/2}. \tag{6.16}$$

The important consequence in establishing the existence of an upper bound for the stress-intensity K_∞ is that it is possible to define a characteristic crack length Λ above which an increase in true crack size a in the component produces no significant increase in stress intensity, see Figure 6.13(d). The value for the characteristic crack length is obtained from equations (6.15) and (6.16), taking

$$K_\infty = \sigma_0(\pi\Lambda)^{1/2}. \tag{6.17a}$$

Then Λ is given by

$$\pi\Lambda = (t_A/G_A)E_P t_P\beta. \tag{6.17b}$$

Analytical results have also been obtained for the situation where the shear yield stress, τ_y for the adhesive is exceeded. Assuming elastic-perfectly plastic behaviour it can be shown that

$$K'_\infty = \tau_y[(t_A E_P)/(3G_A t_P\beta)^{1/2}[1 + 2(F/F_y)^3]^{1/3}. \tag{6.17c}$$

6.4.2 Characterisation studies on an overlap-joint

Our experimental study had the following main objectives:

(i) To assess the validity and accuracy of the analytical theory, based on the one-dimensional analysis of the overlap-joint, by comparing measured values of δ with theoretical predictions,

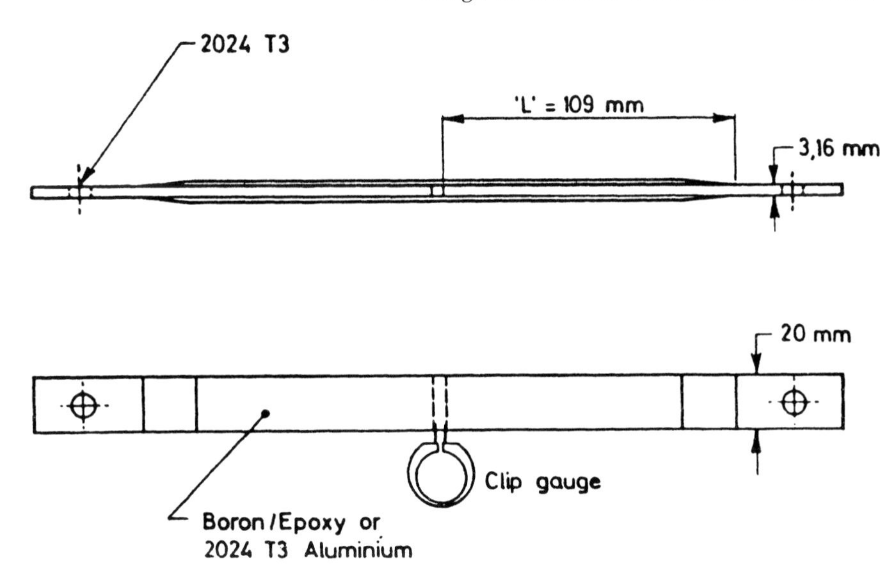

Figure 6.15. Elemental joint specimen (DOFS) showing position of clip-gauge used to measure the gap displacement Δ.

(ii) to characterise the nature and extent of deviations from elastic behaviour due to plasticity, creep or visco-elasticity at load levels representative of those expected in a typical repair.

Experiments [23] were performed on the specimens illustrated in Figure 6.15. The gap displacement Δ (Figure 6.15) was measured by a (strain gauge instrumented) clip-gauge which provided an experimental value for δ through the relationship.

$$\Delta = he_R + \delta \tag{6.18}$$

where e_R, the strain in the reinforcement over the gap, is obtained from equation (6.12).

In each joint the inner adherends were 2024-T3 clad aluminium, Table 6.7 lists the types of specimen studies and the focus of interest in each case.

Table 6.7. Details of joint specimens evaluated.

Specimen type	Outer adherend	Adhesive	Nominal adhesive thickness t_A (mm)	Comments
A	b/ep $t_R = 0.51$ mm	AF 126	0.13	Adhesive used in our repair work.
B	2024–T3 $t_R = 1.58$ mm	AF 126	0.13 To 0.53	To compare metallic with composite reinforcement and to obtain measurements over a range of adhesive thickness.
C	b/ep $t_R = 0.51$ mm	EC2216	0.17	Soft adhesive to simulate behaviour expected from AF 126 at temperatures above 50°C.

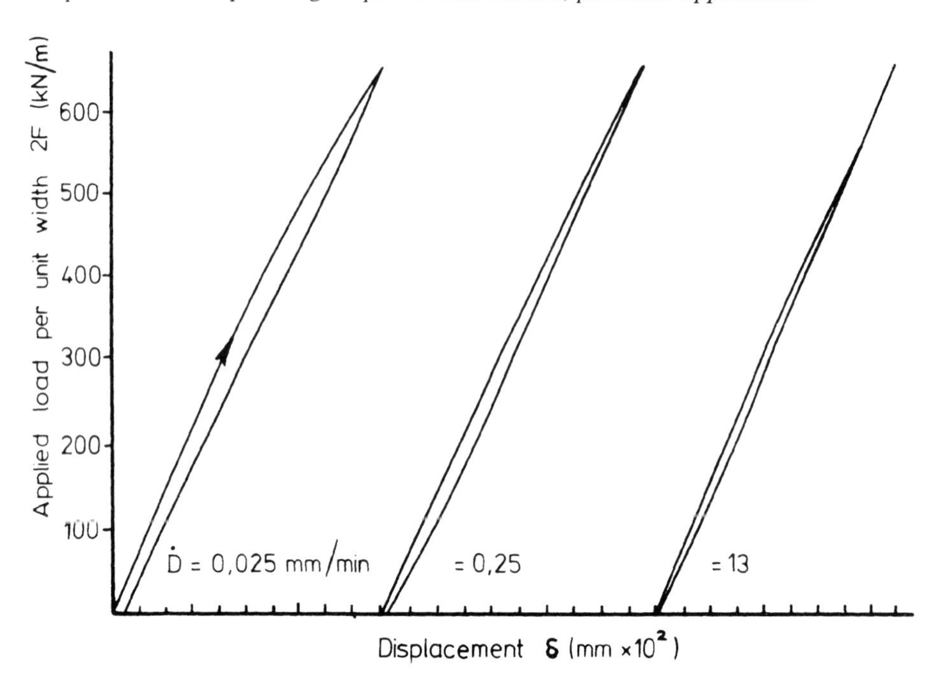

Figure 6.16. Plot of $2F$ versus δ at various cross head speeds ($\dot{D}$) for a type A specimen.

The main findings were as follows:

(a) The variation of δ with applied load intensity $2F$ was approximately linear for the type A and type B specimens within the load range studied, Figure 6.16. However, there was a small but definite hysteresis. The width of the loops was found to increase with decreasing loading rate, indicating a viscoelastic origin of the hysteresis.

(b) Viscoelastic hysteresis was found, as expected, to be much more pronounced with the type C specimen than for A and B types. Figure 6.17 shows this behaviour and also shows a marked drift or creep in δ during the first few cycles of loading. This implies a corresponding decrease in the efficiency of reinforcement.

(c) For the type B specimens (aluminium alloy reinforcement and epoxy-nitrile adhesive) good agreement between observed and theoretically (from the one-dimensional analysis) predicted values of δ could be obtained over a range of adhesive thicknesses, as shown by the results listed in Table 6.8 provided that a shear modulus G_A of 0.54 GPa was assumed for the adhesive, in accordance with results obtained with short overlap shear tests [24]. No allowance for yielding of the adhesive was made here since little effect due to yielding was predicted [23] at the stress level employed in these tests. However, the good agreement should not be over-emphasized since experimental errors mainly due to variations in adhesive thickness (difficult to avoid in practice), can be significant.

(d) For the type A specimens (b/ep reinforcement and epoxy nitrile adhesive) δ values were generally slightly larger than for the type B specimens, as given in Table 6.8, even though the product $E_R t_R$ was nominally similar for each reinforcement.

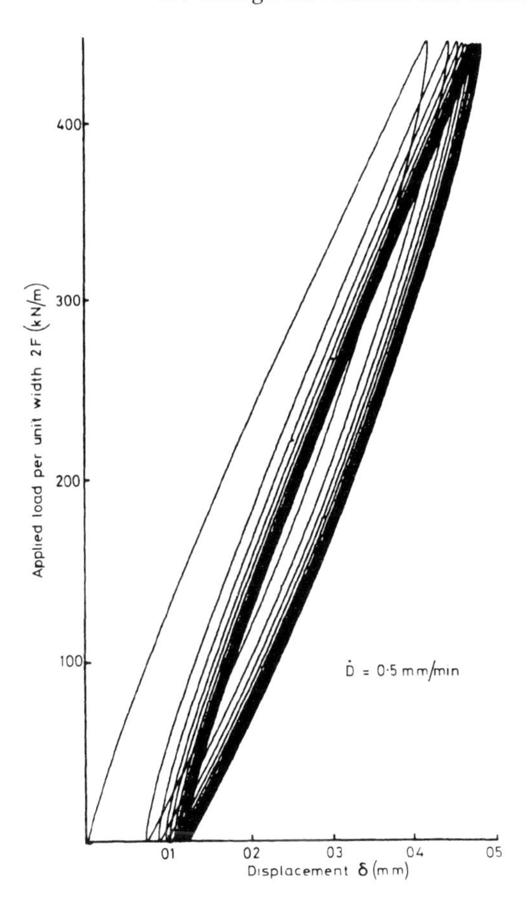

Figure 6.17. Plot of load intensity $2F$ versus δ during cyclic loading to a steady state for a type C specimen.

Table 6.8. Experimental results and theoretical values of δ at $2F = 680\,\text{kN/m}$ for the overlap joint specimens.

Specimen number	Experiment			δ Displacement (μm)			
	t_A (mm)	h (mm)	Δ (μm)	(a)	(b)	(c)	(d)
A1	0.12	4.44	43	29	22	28	35
A2	0.13	4.52	44	30	23	30	34
A3	0.13	4.54	42	28	23	30	34
B1	0.17	4.32	39	27	24	–	–
B2	0.20	4.52	43	30	27	–	–
B3	0.14	4.34	37	25	23	–	–
B4	0.33	4.46	46	32	35	–	–
B5	0.38	4.57	48	35	38	–	–
B6	0.53	4.56	55	42	45	–	–

Notes: (a) Experimental result. (b) No correction for shear lag, taking $G_A = 0.54\,\text{GPa}$. (c) Correction for residual-stress $\Delta\delta$ shear lag, using equation (6.19). (d) Correction for residual stress σ_T using equations (6.9a) and (6.10b).

Two possible explanations can be given for this, either alone capable of providing sufficient correction.

The first possibility is that, shear-lag effects in the b/ep reinforcement, due to its relatively low shear modulus, Table 6.1, increase the characteristic load transfer length (β^{-1}) and thus increases δ. A simple correction to β, assuming a linear shear-stress variation across the thickness [25] can be made by replacing (t_A/G_A) in equation (6.1c) with the effective value

$$(t_A/G_A)_{\mathrm{eff}} = (t_A/G_A + t_P/2G_P + 3t_R/8G_R) \qquad (6.19)$$

where G_P and G_R are, respectively, the shear moduli in the through-thickness plane for the metal and composite component respectively. When this correction is made for the theoretical δ value (as given in Table 6.8), good agreement is obtained with the experimentally determined δ for these specimens.

The second possible explanation is concerned with the influence of the residual stress σ_T which arises due to thermal expansion coefficient mismatch effects. It was shown earlier that, if σ_T is significant compared to σ^{∞}, δ will be increased according the equations (6.8) or (6.9a) and equation (6.10b). However, in the experiments only the difference in displacement $\Delta\delta$ was measured (since the displacement δ_T is present prior to loading), thus the influence of σ_T would only be apparent if $F > F_y$ so that the displacement was as given by equation (6.9a); this was found to be the case in the experiments, based on the apparent shear yield stress for the adhesive $\tau_y = 30$ MPa, taken from reference [24]. Using equations (6.9a) and (6.10b) for this value of τ_y the theoretical values of $\Delta\delta$ were calculated. Table 6.8 shows that allowance for σ_T in this way results in over-correction of the experimental measurements.

Limitations of the one-dimensional analysis and joint specimen

A finite-element analysis [26] was undertaken to study the importance of two-dimensional effects; only elastic behaviour was considered. Table 6.9 provides the main result of the analysis.

(i) Concerning firstly displacements Δ and δ, consideration of Tables 6.9a and 6.8 shows that excellent agreement is obtained between experimental measurements (bearing in mind the possible errors referred to earlier), the finite-element analysis and the one-dimensional analysis. Errors due to through-thickness variations shown by comparison with Δ' and δ', are significant but not excessive.

(ii) The nominal shear strains γ_A and γ'_A derived from the finite-element results for displacements δ or δ' (using equation 6.7) are compared with the finite-element results for true shear strain (τ_A/G_A) in Table 6.9a. This comparison shows that the nominal shear strain γ'_A, if based on experimental measurement of displacement in the overlap specimens close to the inner adherend surface, is expected to provide a good estimate of the true shear strain.

(iii) The variation in shear stress (τ_A) with gap size (h), indicated in Table 6.9b, appears to be quite small. Thus the overlap specimen is quite representative of a cracked specimen $(h = 0)$ in this respect.

(iv) Through-thickness stresses σ_z in the adhesive are shown, Table 6.9b, to be negative (anti-peel) and to rapidly decrease with decreasing h. Thus the overlap

Table 6.9. Results based on the finite-element analysis of the joint specimen given in reference [26]. A force intensity ($2F$) of 680 kN/m, adhesive thickness t_A of 0.13 mm and adhesive shear modulus G_A of 0.54 GPa is assumed. All other properties are as listed in Table 6.1.

(a)

Displacements (μm)								Shear Strain			
Experiment		Analysis		Finite-element analysis				Finite-element analysis			
Δ	δ	Δ	δ	Δ	Δ'	δ	δ'	γ_A	γ'_A	τ_A/G_A	
44	30	–	23	41	37	26	21	0.100	0.084	0.073	Specimen A
37	25	–	23	38	34	24	20	0.092	0.076	0.071	Specimen B

Note: (i) The $\Delta\delta$ and γ_A F-E values are based on the centerline of the joint; the Δ', δ', $\gamma_A{}'$ are F-E values based on the edge of the inner adherend. (ii) Experimental results and analysis are taken from Table 6.8.

(b)

Specimen A			Specimen B		
h(mm)	σ_zMpa	τ_A(MPa)	h(mm)	σ_z(MPa)	τ_A(MPa)
8	−30	39	8	−26	38
4	−27	40	4	−21	38
2	−23	41	2	−18	39
1	−20	42	1	−13	39
0.25	−17	43	0.25	−10	41
0	−15	44	0	−8	42

Note: (i) σ_z is through thickness stress in adhesive layer. (ii) h is the length of the gap, see Figure 6.14.

(c)

Stress distribution through outer adherend						
h = 4 mm			h = 0 mm			
Outside surface (MPa)	Inside surface (MPa)	Stress ratio	Outside surface (MPa)	Inside surface (MPa)	Stress ratio	
627	786	1.3	492	1205	2.5	Specimen A
374	546	1.5	276	747	2.7	Specimen B

specimen is only completely representative of a cracked specimen when h is made close to zero.

(v) Concerning the through-thickness stress variations in the outer adherend, Table 6.9c shows that significant strain concentrations arise at the bonded surface particularly when $h = 0$. This important aspect cannot be predicted by the one-dimensional theory and some allowance based on the finite-element analysis will be required in assessing the allowable strain e_R in the reinforcement. Joint specimens to investigate the allowable e_R experimentally should therefore have small h.

The main conclusion drawn from above studies on the joint specimens is that with the epoxy-nitrile adhesive AF126 and the aluminium alloy reinforcement, the simple one dimensional analysis is applicable for estimating δ and hence γ_A and

K_∞. In the case of joints with b/ep reinforcement, correction must be made for shear-lag and for residual-stress effects. The one-dimensional analysis cannot provide an estimate of the peak strain in the reinforcement; it only gives the average value. More complex analysis is required where marked plasticity or viscoelastic effects become significant, as was found with the epoxy-paste adhesive and would be expected for the epoxy-nitrile adhesive at elevated temperature.

The experimental technique based on the overlap joint is considered to be a very useful means of reproducing the conditions in the adhesive layer of a patched specimen. However, this approach only provides representative values for e_R, the strain in the reinforcement, when the gap length h is close to zero.

6.4.3 Cyclic loading studies on an overlap-joint

The aim of this preliminary study was to develop procedures for obtaining conservative values for the patching allowables – particularly for fatigue properties of the adhesive. A plausible criterion for fatigue damage in the adhesive (based on experience with metallic materials) is its plastic shear strain range, under cyclic loading. However, since the plastic strain range is difficult to assess accurately, the nominal total shear strain range $\Delta\gamma_A$ was used in our study as the damage criterion – based on $\Delta\delta$ the range of displacement under cyclic zero/tension force intensity ($2F$). In fact, both the total and the plastic shear strain range have been successfully employed [27] as criteria for cyclic damage in plastics.

Cyclic loading experiments were performed on joint specimens with similar configuration to that depicted in Figure 6.15, except that the thickness of the adherends was doubled to allow higher load intensity. Specimens were bonded with adhesive AF126 at 120°C, using either b/ep or 2024 T3 outer adherends; the pre-bonding surface-treatment was by the PANTA process. In the test the amplitude of $2F$ was kept constant (frequency 0.6 Hz) and the gap displacement Δ was plotted as a function of cycles; Figure 6.18 is a typical experimental record.

Figure 6.18 shows that (i) substantial creep of the minimum gap displacement (Δ_c) occurs during cycling, and (ii) the difference between the minimum and maximum displacements (Δ and Δ_c) increases during cycling. In general, the creep displacement Δ_c appears to be fully recoverable following unloading of the specimen. It therefore appears that the Δ_c curve is essentially a static stress-relaxation phenomena, similar to that discussed in section 6.2.1. However, the increase in $(\Delta - \Delta_c)$ with increasing numbers of cycles appears to be attributable to damage or deterioration in the adhesive system due to the cyclic loading.

Fatigue damage

The most likely forms of damage are (i) gross cracking or disbonding in the adhesive layer, – and/or (ii) crazing (or some more subtle deterioration) of the polymer structure resulting in a reduction in the effective shear modulus of the adhesive (G_A).

For the case of cracking or disbonding of the adhesive system assuming no change in shear modulus G_A, using equation (6.18), we have

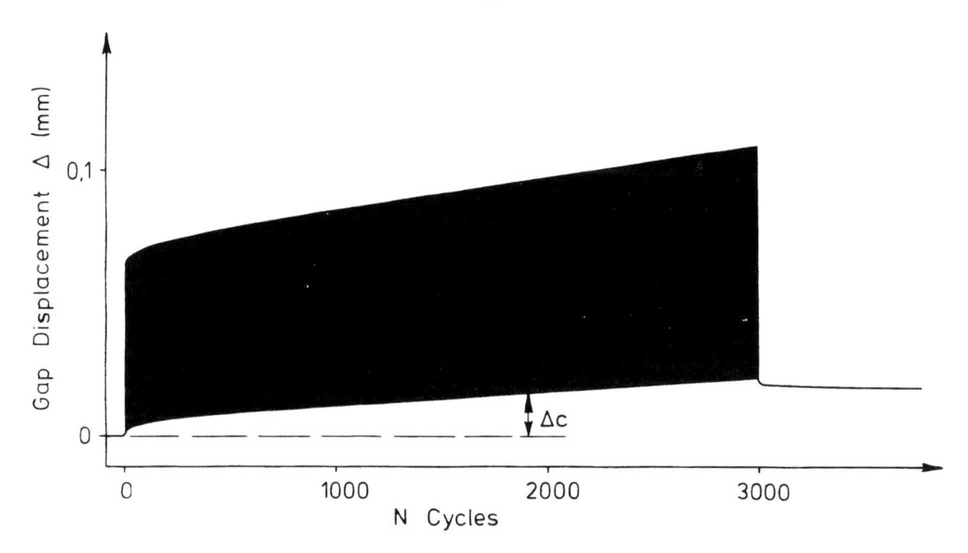

Figure 6.18. Plot of gap displacement (Δ) versus cycles for an overlap joint with b/ep outer adherends, $2F$ equals 1335 kN/m.

$$\Delta = (h + 2a)e_R + \delta \tag{6.20a}$$

and

$$\delta = \Delta\delta + \delta_c \tag{6.20b}$$

where 'a' is the effective crack length, δ_c is the displacement of the adhesive due to creep (equal to Δ_c) and $\Delta\delta$ is the range of displacement of the adhesive. The apparent shear strain range is then given by $\Delta\gamma_A = \Delta\delta/2t_A$. The effective crack length can be obtained from

$$2a = [(\Delta - \Delta_c)_N - \Delta_0]/e_R \tag{6.21}$$

where the subscripts N and zero refer to measurements of Δ after N cycles and at the beginning of the test respectively. Data in the form used in Figure 6.18 for Δ versus N can be converted to plots of δ versus N through equation 6.18, initially making no allowance for crack or damage growth. Typical results for the two types of joint specimen are plotted in Figures 6.19(a) and (b).

In order to assess the nominal values of $\Delta\delta$ and thus $\Delta\gamma_A$ it is necessary to use equations (6.20) and (6.21). These values are indicated on Figure 6.19. Also shown as an inset on this Figure are approximate estimates of $\Delta\delta$ and $\Delta\gamma_A$ from equations (6.9) and (6.10), assuming τ_y, the yield stress of the adhesive, equals 30 MPa. If now the nominal crack length a (from equation (6.21)) is plotted against N, as given in Figure 6.20, linear crack growth relationships result, against which the nominal values of $\Delta\gamma_A$ can be indicated; this is the design data required. Straight line crack growth relationships are expected since a constant energy release rate G is predicted for a crack or disbond growing in the adhesive layer from the gap region in these specimens.

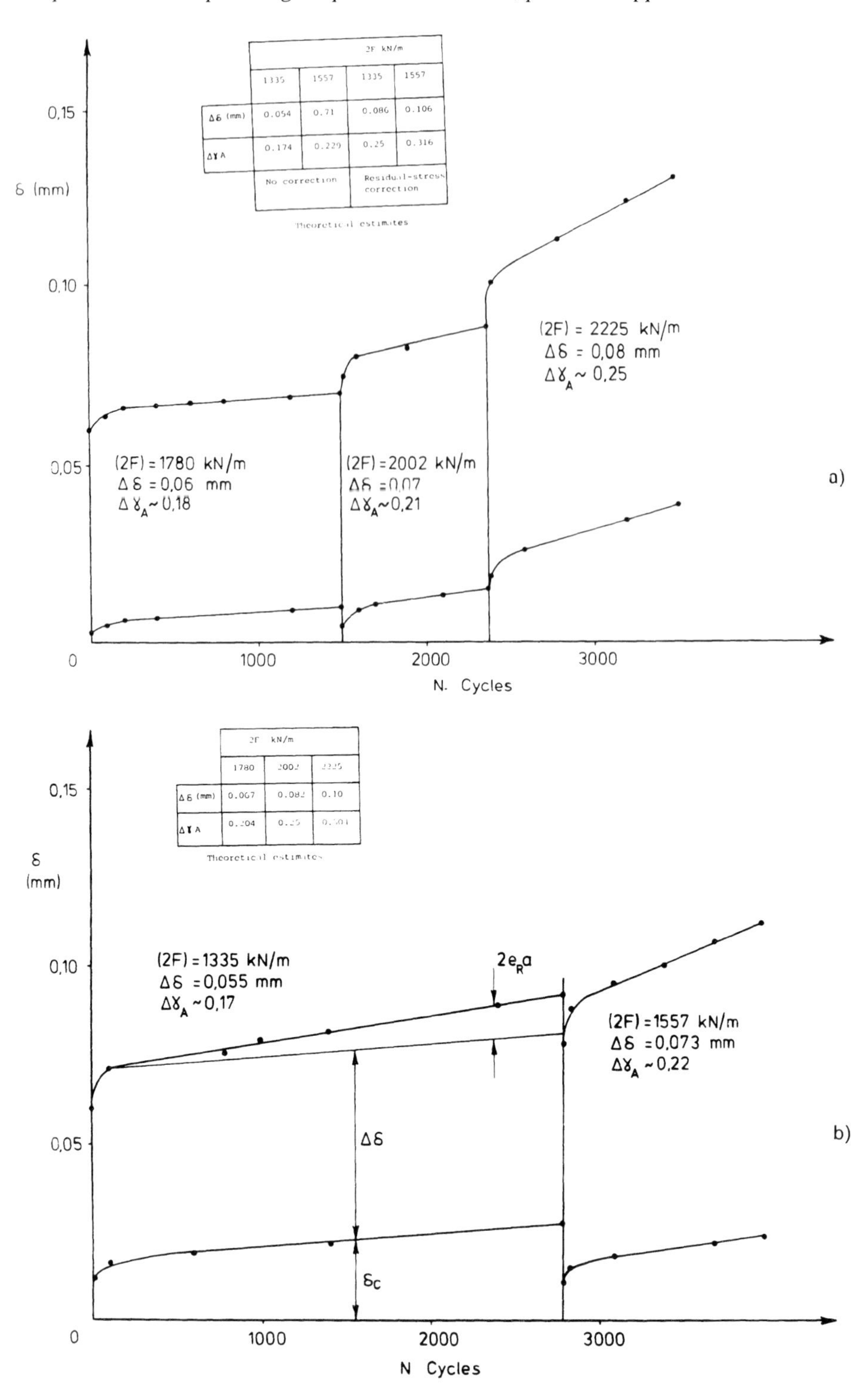

Figure 6.19. Plots of nominal δ versus N for specimens (a) with b/ep outer adherends and (b) 2024 T3 adherends; in both cases $t_A \sim 0.17$ mm. Inset tables give theoretical values based on equations (6.9) and (6.10).

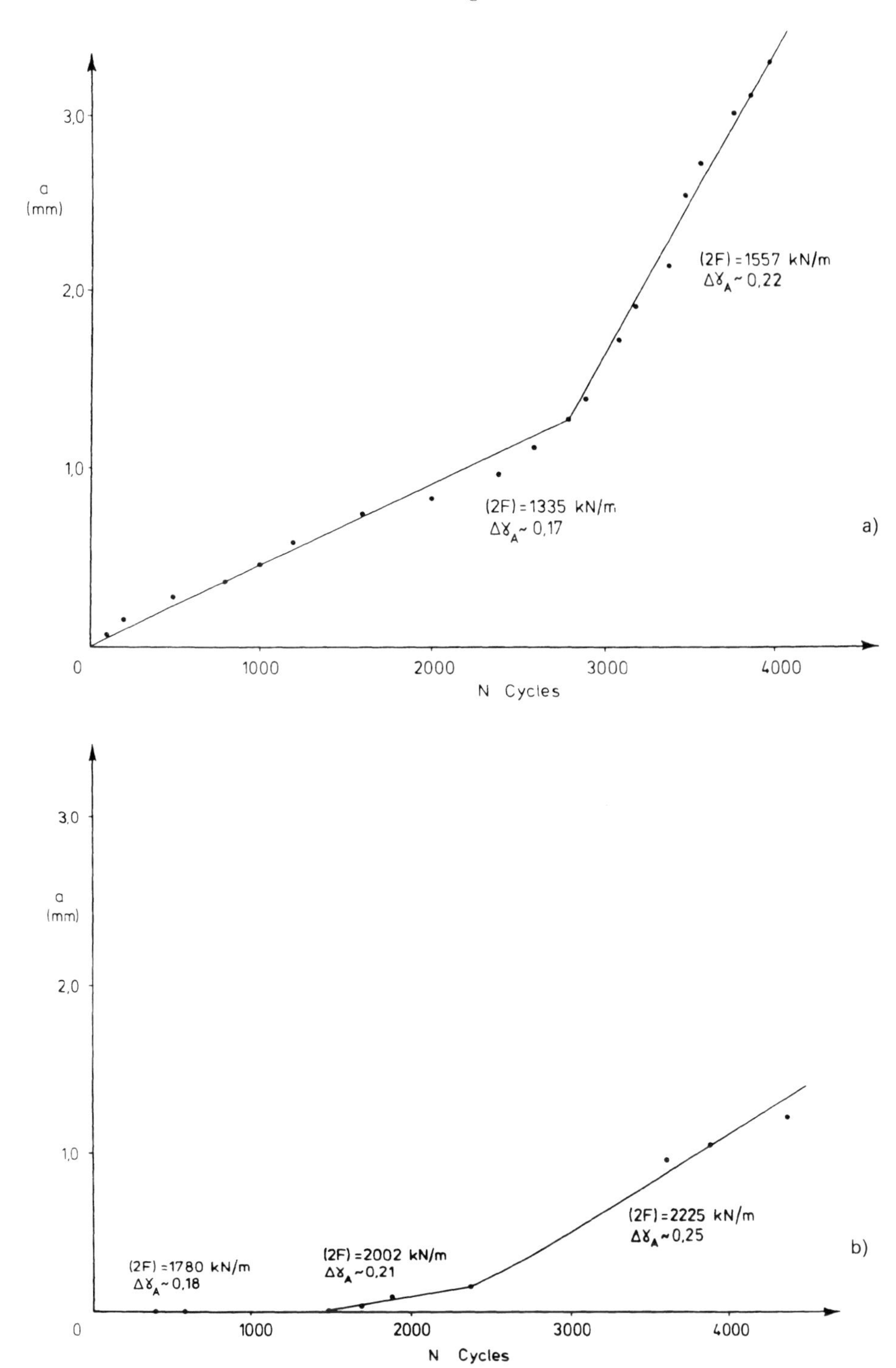

Figure 6.20. Plots of apparent crack size a in the adhesive layer versus N for specimens with outer adherends of (a) b/ep and (b) 2024 T3; data taken from Figure 6.17(a) and (b).

The inferior performance of the b/ep reinforced specimen compared to the 2024 T3 reinforced specimen (Figure 6.20(a)) possibly results from the presence of high residual stress. It should be noted, however, that (as discussed in section 6.3) residual stresses would be substantially lower in a patched component than in a simple overlap joint.

Unfortunately, the simple interpretation of behaviour in terms of cracking in the adhesive layer assumed above may be invalid since cracking was not always observed metallographically in these specimens. It thus appears that damage should be interpreted in terms of the reduced G_A. The absence of cracking in some experiments is in contrast to the observations of cracking noted in section 6.2.1 and may be related to (i) the large difference in cyclic loading frequencies employed in the two studies – 0.6 Hz here and 50 Hz previously; for example, large scale relaxation would not arise at 50 Hz, or (ii) the presence of the compressive stress σ_z (see Table 6.10b) which may inhibit cracking.

Despite the absence of macro cracking in some specimens, the concept of effective crack size is useful for design purposes, particularly in view of the observed linear behaviour of 'a' versus N. An alternative idealisation is that 'a' represents an ineffective zone (a region of zero load transfer) rather than a true crack. In any event, a knowledge of the allowable $\Delta\gamma_A$ for acceptable levels of da/dN (leaving open the interpretation of a) is a most useful approach for repair design. The influence of fatigue damage and stress relaxation in the adhesive on K_∞, the upper bound for the stress-intensity range experienced by the cracked component, follows from equation (6.15a) and the experimental value of δ. Then, at least to a first approximation

$$K_\infty = (E_P F\delta/2(1 + S)t_P)^{1/2}. \tag{6.22}$$

Thus the curves of δ versus N such as shown in Figure 6.19 are readily converted to K_∞ versus N.

6.5 A preliminary design approach

This section describes an approach to patch design, based largely on the analysis outlined in section 6.3. Repairs in critical or complex situations would generally have to undergo final design by the finite element procedure. The main aim of the design approach is to provide a feasibility assessment of repairability of a component with a fatigue crack, and if repair is feasible, to provide estimates of the following:

(i) minimum required patch thickness,
(ii) minimum required overlap length,
(iii) conservative value for the reduction in stress intensity,
(iv) expected durability of the repair scheme.

Only the situation where the crack is completely covered by the reinforcement is considered here. The reinforcement may cover the crack on one side or both sides of the component. However, where reinforcement is only on one side, it is assumed here that secondary bending effects, such as produced by displacement of the

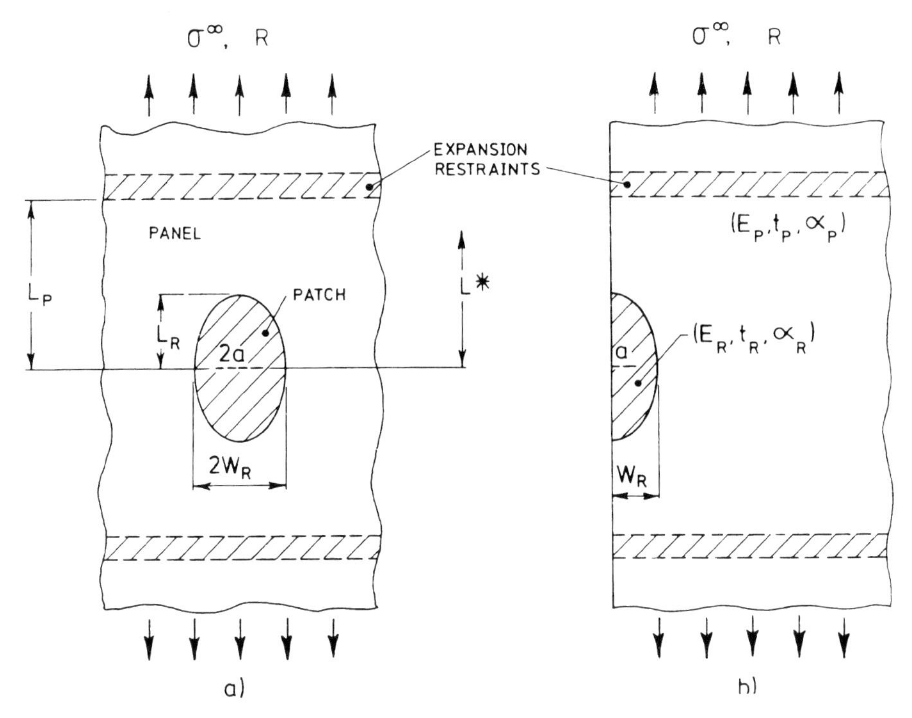

Figure 6.21. Schematic of patching situation assumed in the analysis with (a) center crack, and (b) edge crack showing the position of restraints to thermal expansion.

neutral axis by the patch, are reacted out by the supporting structure. Corrections for bending effects are given in Chapters 4 and 5.

It is generally desired to employ the thinnest patch feasible for several reasons, including to (a) minimise the residual stress problems indicated in Figure 6.9, (b) maintain aerodynamic acceptability, for example to minimise disturbance to the airflow when repairs are made to an external surface, (c) minimise balance problems; for example, when repairs are made to a control surface, and (d) comply with installation restraints, for example, not to exceed available fastener lengths when fasteners must pass through the patch for system requirements, or to maintain clearance between moving surfaces.

Design input

In an idealised fatigue cracking situation, where the cracked component is modelled as a large panel, either center or edge-notched, as depicted in Figure 6.21, the following relevant design information is available or can be estimated, initially concerning the cracked component:

(i) The thickness t_p and modulus E_p of the metallic component. Note that t_p (as defined here) is the thickness of the metallic component if it is patched on one side only or half the thickness if patches are applied on each side of the component.

(ii) Length of the *longest* available (allowable) overlap length, L^* perpendicular to the crack that can be covered by the patch and the proposed patch width $2W_R$.

(iii) The magnitude of the peak cyclic stresses normal to the crack σ^∞ and parallel to the crack σ_x^∞ and the ratio R of σ^∞ to the minimum normal stress.

(iv) The coefficient of thermal expansion for the metallic component α_P and the degree of restraint by the surrounding structure based on the distance $2L_p$ in the fibre direction between rigid connections (for example, to a substructure), see Figure 6.21.

(v) The operating temperature and environment.

(vi) The crack size a.

In order to assess the significance or effectiveness of repair, it is desirable to have representative data relating to the rate of crack growth as a function of cyclic stress-intensity range ΔK, including crack-growth retardation effects and influence of the stress ratio R.

Concerning the patch, the design information required is:

(i) Tensile and shear modulus, E_R and G_R respectively.

(ii) Thickness per ply, T_R.

(iii) Allowable tensile strain, e_R^* (allowing for any strain concentration).

(iv) The coefficient of expansion α_R.

Finally, concerning the adhesive, the design information required is:

(i) Thickness, t_A,

(ii) Effective shear modulus G_A,

(iii) Effective yield shear stress τ_y,

(iv) The allowable cyclic shear strain $\Delta\gamma_A^*$, for various levels of durability,

(v) The adhesive cure temperature, T_A.

The parameters G_A, $\Delta\gamma_A^*$, τ_y and, to a lesser extent, G_R are sensitive to the environmental temperature and humidity (if humidity is not to be excluded) and to the rate of loading.

Design procedure-minimum thickness patch

The logic for the design procedure, shown in flow chart form in Figure 6.22, is based on comparison of:

(i) The computed overlap length L_R with the allowable length L^*.

(ii) The computed peak strain in the patch e_R with the allowable e_R^*.

(iii) The computed shear strain range $\Delta\gamma_A$ in the adhesive with the allowable $\Delta\gamma_A^*$ as the patch is increased in thickness one ply at a time. Increasing patch thickness (as shown later), increases L_R but, as can be seen from the theoretical results given in Section 6.4.1, reduces both e_R and $\Delta\gamma_A$. This procedure results in specification of the thinnest patch system which, under the applied loads, will not exceed the available length and thus will not fail under the service loads. The peak shear strain e_R would generally be chosen as a criterion for damage in the reinforcement, rather than the strain range, since the unidirectional fibre composite is not usually prone to reduction in residual strength when subjected to cyclic tension loading. The level chosen for the allowable peak shear strain e_R^* can thus be similar to the allowable static strain value – allowing for strain concentration effects, see section 6.4.2. It is relevant to note here, however, that due to residual stress (see section 6.3) the reinforcement will generally be under compression which will enhance its tensile strain capacity [17].

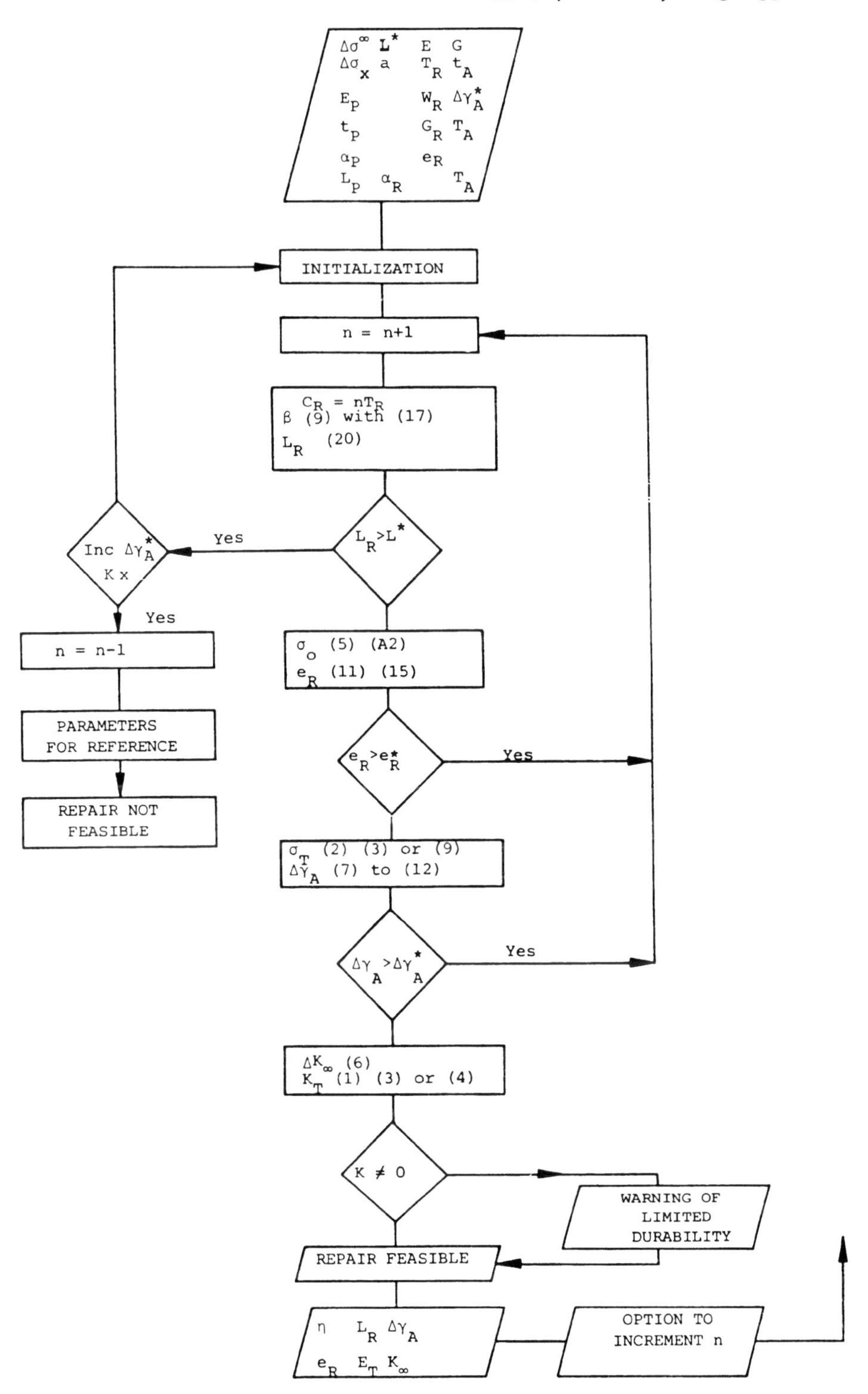

Figure 6.22. Flow diagram for patch analysis.

The level chosen for the allowable shear strain range in the adhesive $\Delta\gamma_A^*$ is assumed to determine the fatigue life or durability of the patch system. As indicated in Figure 6.22 several increments in $\Delta\gamma_A^*$ above the value initially taken are allowed if the required overlap length exceeds the length available. The input $\Delta\gamma_A^*$ is initially chosen to provide the required durability with a large margin of safety; accordingly, increasing levels of $\Delta\gamma_A^*$ are associated with decreasing durabilities. The durability information would be based on fatigue tests on candidate adhesives, as described in Section 6.4.3. The choice of $\Delta\gamma_A$ as the main design criteria is based on the assumption that the failure mode is cohesive, due to fatigue cracking of the adhesive layer. However, should adhesive (debonding) failure occur, the nominal $\Delta\gamma_A$ for the adhesive would probably remain as a useful indicator of the severity of the loading of the adhesive system.

Stress-intensity considerations – patching efficiency

No attempt is made in the design approach (above) to design to a specified stress intensity or stress intensity factor range ΔK_∞. This is because of (i) the uncertainties in the estimate of ΔK_∞, (ii) the difficulties of accounting for the stress-ratio effects, (iii) the difficulty in allowing for overload or crack-retardation effects, and (iv) the large errors in the prediction of crack growth rates due to materials variability and environmental influences. Moreover the estimate of K_∞ rapidly becomes insensitive to increasing patch thickness.

In an arbitrary cracking configuration, the value of ΔK_∞ for a center-cracked panel may have limited significance. It is, however, assumed that $\Delta K_\infty/\Delta K_a$ the stress-intensity ratio (where ΔK_a is the stress-intensity range in an unpatched center cracked component) provides a useful indicator of patching efficiency, given by

$$\Delta K_\infty/\Delta K_a = (\Delta\sigma_0/\Delta\sigma^\infty)(\Lambda/a)^{1/2} \tag{6.23}$$

As discussed in Section 6.3 the presence of a residual-stress intensity K_T must also be considered. In the design program the residual-stress σ_T is calculated from an estimate of the effective thermal expansion coefficient of the metallic component, α'_P. To find σ_T, an assessment is required of the degree of restraint to expansion during the patch application procedure, along the lines discussed in Section 6.3.

On the basis of the applicability of equation (6.4) for the center-cracked panel, the maximum expected value of K_T in an arbitrary situation is given by a general relationship such as

$$K_T = \sigma_T(\pi a)^{1/2}Y \tag{6.24}$$

where Y is a geometric factor, characteristic of the particular cracking problem and is taken as unity for a center-cracked panel.

Most of the relationships required in the design program have been presented earlier, particularly in Section 6.3.1, other than the relationships for L_R the overlap length required which are given below.

Overlap length L_R

An estimate for the overlap length L_R required for the patch can be obtained using the approach for joint design taken in reference [28], again modelling the patched region as an overlap joint, Figure 6.14. If the degree of yielding of the adhesive is fairly small, a conservative estimate for L_R is given by

$$L_R = 6/\beta + \text{length of taper} \tag{6.25}$$

where β is given by equations (6.1c) and (6.19). The overlap length $6/\beta$ the minimum length required to fully develop the region of low shear stress at the center of each side of the joint. This central region serves to (i) minimise stress-relaxation in the patch under static loads, (ii) provide allowance for manufacturing defects, such as voids and disbonds, and (iii) allow for some service deterioration. To minimise peel stresses, the outer edges of the patch must be tapered; we generally employ a taper angle of about 5°. Conservatively, this length is added to the elastic transfer length.

Some results of the application of the design procedure

Repair design studies [29] were undertaken on metal panels having the configuration illustrated in Figure 6.21. The panels are made either from aluminium or titanium alloy and the repair system consists of b/ep unidirectional patches bonded to one side of the panel with AF126 adhesive – the panels are assumed to be restrained against bending. The level of restraint of the panels against thermal expansion, when curing the adhesive at 120°C, is determined by the separation L_p of the strong connection, shown shaded. However, in some cases, the condition of no restraint (free edges) is assumed.

Table 6.10 lists representative input data and corresponding output data including patch thickness, in terms of n – which is the minimum number of plies required to avoid adhesive or patch failure. The minimum allowable shear strain range $\Delta\gamma_A^*$ is established by fatigue testing of the overlap joint specimen and, for adhesive AF126, is taken here to be 0.18. The value for $\Delta\gamma_A^*$ is allowed to increase from the minimum value of 0.18 to a maximum of 0.25, if required, with the corresponding loss of durability noted as part of the output of the program. There are two cases where repair is not feasible; in case 5, the maximum allowable $\Delta\gamma_A^*$ is exceeded and in case 7, the allowable patch strain e_R^* is exceeded. Note, this exercise makes no allowance for strain concentration in the reinforcement.

The thickness of the patch can be increased to reduce further the stress intensity and the strain levels as long as the increase does not result in the required patch length L_R exceeding the available length L^*. Table 6.11 shows the effect of increases in n for some of the cases. It is interesting to note that (as mentioned earlier) the reduction in K_∞ or K'_∞ with increasing patch thickness is not very marked.

Case 1 approximates to the case of the edge-notched panel tests described in section 6.6 in which an aluminium alloy panel 3.14 mm thick is patched without restraint (free edges) during adhesive cure.

The design output calls for a seven ply patch 114 mm long. The upper bound stress-intensity ΔK_∞ calculated from equations 6.17a and b is about a third of the unpatched value ΔKa. However the upper bound stress-intensity $\Delta K'_\infty$ calculated from equation 6.17c, which allows for plastic yielding in the adhesive, is about twice ΔK_∞. The difference between ΔK_∞ and $\Delta K'_\infty$ arises mainly because the large

Table 6.10. Input data and output results for minimum patch thickness, based on the design procedure.

	Case			(Aluminium alloy)				(Titanium alloy)		
Input	1	2	3	4	5	6	7	8	9	10
σ^{∞} (MPa)	138	→	→	216	138	→	→	→	→	→
R	0.1	→	→	→	→	→	→	→	→	→
E_P (MPa)	72	→	→	→	→	→	→	110	→	→
t_P (mm)	3.14	→	→	→	7	3.14	→	→	→	→
α_P °C^{-1}x10	2.4	→	→	→	→	→	→	9	→	→
L_P (min)	FE	∞	100	→	→	→	→	→	∞	FE
L* (mm)	80	→	→	→	→	→	40	80	→	→
a (mm)	25	→	→	→	→	→	→	→	→	→
E_R (GPa)	200	→	→	→	→	→	→	→	→	→
T_R (mm)	0.13	→	→	→	→	→	→	→	→	→
W_R (mm)	80	→	→	→	→	→	40	80	→	→
G_R GPa)	10	→	→	→	→	→	→	→	→	→
$e_R^* \times 10^{-3}$	5.0	→	→	→	→	→	→	→	→	→
α_R °Cx10^{-6}	4	→	→	→	→	→	→	→	→	→
G_A (GPa)	0.54	→	→	→	→	→	→	→	→	→
t_A (mm)	0.19	→	→	→	→	→	→	→	→	→
$\Delta\gamma_A^*$	0.18	→	→	→	→	→	→	→	→	→
τ_y (MPa)	30	→	→	→	→	→	→	→	→	→
T_A °C	120	→	→	→	→	20	120	→	→	→
Output	1	2	3	4	5	6	7	8	9	10
n	7	5	5	9	9	5	3	4	4	5
L (mm)	57	48	48	65	76	48	37	44	44	49
$\Delta\gamma_A$	0.16	0.16	0.13	0.17	0.28	0.14	0.22	0.16	0.17	0.16
$e_R \times 10^{-3}$	3	4.2	4.2	4.2		4.2	6.5	4.8	4.7	3.9
σ_T (MPa)	67	21	-4	-6	NO REPAIR	0	NO REPAIR	-7	1.2	15
ΔK_∞ (MNm$^{-3/2}$)	12.5	14	14	18		14		19	19	18
$\Delta K'_\infty$ "	25	22	18	22		18.7		24	25	27
ΔK_a "	38.7	→	→	60.5		→	→	→	→	→

Notes: (a) Stress-intensity K_∞ is obtained from equation (6.17a,b). (b) Stress-intensity K'_∞ is obtained from equation (6.17c) with allowance for stress σ_T. (c) F-E signifies free-edge. (d) ∞ signifies edge clamped at infinity.

thermal stress σ_T reduces the effective yield stress in the adhesive. It should be noted that, for most of the other cases σ_T is small or even negative, due to restraint during patching, so that the value of ΔK_∞ is similar to $\Delta K'_\infty$.

Finally, Table 6.10 shows that the minimum thickness patch for cases 8, 9 and 10, for a titanium alloy panel, is less than for the cases based on an aluminium alloy panel of the same thickness. This conclusion of course does not imply that the patch reinforcement is more effective in reducing K for the titanium panel, as shown in Table 6.11 the reverse is true, as expected.

6.6 Crack propagation behaviour in patched specimen

This section describes work to evaluate the fatigue performance of patched specimens and then compare the results with theoretical expectations.

Effective patching is expected to have two major beneficial effects in reducing crack growth. It should (i) retard the reinitiation of the crack, and (ii) reduce the rate of crack growth once growth resumes. Both effects result from the reduction in stress intensity range (following patching) from ΔK_a to ΔK_R, the effective value which should be less than ΔK_∞ maximum predicted value.

Retardation is usually associated with application of an overload during cyclic loading of a cracked specimen [30]; it should also occur when the effective cyclic load (or stress-intensity range) is reduced by patching. Essentially, retardation is associated with the formation of a plastic zone at the crack tip during loading, the zone acts as an oversize inclusion, resulting in a compressive residual stress from the surrounding elastic material when the loads are removed. Thus when a high load occurs, it produces a relatively large plastic zone and, on unloading, a correspondingly large residual stress. If a lower load is now applied, the residual stress acts on the crack and crack tip (a) delaying the opening of the crack, thereby reducing the effective stress intensity, and (b) delaying the onset of further plasticity at the crack tip. Since, during cyclic loading the damage per cycle is directly related to the size of the plastic zone formed during that cycle, the result of the overload is a marked reduction in crack growth rate, (depending on the relative size of the high load), until the crack grows out of the region (overload plastic zone) influenced by the residual stress. Under variable cyclic loading the degree of retardation following patching will differ from that obtained under constant amplitude cyclic loading and will depend on the load level experienced by the cracked component just prior to patching. However, retardation effects should still play an important role in determining patching performance.

Following patching, the reduced rate of crack growth reflects the reduction in stress intensity range once the retardation effects are past. Further, since, (as previously discussed) patching theory predicts that for total reinforcement the maximum stress intensity range following patching (ΔK_∞) should be independent of crack size, the rate of growth $\mathrm{d}a/\mathrm{d}N$ is expected to be constant, on the basis of the empirical relationship

$$\mathrm{d}a/\mathrm{d}N \propto (\Delta K)^n \tag{6.26}$$

where n is an exponent usually between 3 and 4.

Finally, two potentially important complicating factors must also be considered. The first is the presence, following patching of the residual stress σ_T; the second is the influence of the heat-treatment following patching on the behaviour of the cracked component.

Table 6.11. Output data for increasing patch thickness.

Case	1			2				3				10		
n	9	11	13	7	9	11	13	7	9	11	13	7	9	11
L_R(mm)	65	72	78	57	65	72	79	57	65	72	79	59	67	75
$\Delta\gamma_A$	0.14	0.12	0.11	0.12	0.1	0.09	0.08	0.1	0.09	0.08	0.07	0.12	0.1	0.09
$e_R \times 10^{-3}$	2.7	2.3	2.0	3.2	2.7	2.3	2.0	3.2	2.7	2.3	2.0	2.9	2.4	2.1
σ_T(MPa)	7.3	80	92	2.6	29	32	35	−5	−5.9	−6.4	−6.9	19	22	25
ΔK_∞($MNm^{-1/2}$)	11.4	10.5	9.8	12.5	11.4	10.5	9.8	12.5	11.4	10.5	9.8	16	14.5	13.5
$\Delta K'_\infty$($MNm^{-1/2}$)	25	25	25	20.5	19.8	19.2	18.8	16.6	15.4	14.6	13.9	25	24	24
K_T($MNm^{-1/2}$)	14.5	16	18	5.2	5.7	6.3	6.9	−1	−1.2	−1.3	−1.4	3.67	4.3	4.8

The following experiments on fatigue crack propagation were made to evaluate the above aspects;

(i) Patched specimens with total reinforcement were cyclically stressed to evaluate patching efficiency and compare with theoretical predictions.

(ii) Unpatched specimens were cyclically stressed to evaluate the influence of heat-treatment to conditions representative of those employed to bond the patch, and

(iii) Patched specimens with tip reinforcement were cyclically stressed to compare performance with total reinforcement. Tip reinforcement would be employed in practice only when the main body of the crack was inaccessible.

6.6.1 Patched specimen – total reinforcement

Edge-notched specimen, as depicted inset in Figure 6.23, were subjected to cyclic loading at constant load amplitude. The specimens conformed with the idealised situation illustrated in Figure 6.21(b) and most of the design conditions conformed with the situation assumed for case 1, Table 6.11, in which the patch is assumed to be bonded to one side of the panel with AF126 adhesive at 120°C – with no edge restraint during bonding. In some specimens the patches were bonded with AF126 at 80°C, to evaluate performance with minimum bonding conditions (see section 6.2.2) and in other specimens the patches were bonded with FM300 (by Cyanamid), cured at 175°C, to evaluate performance with an adhesive with a significantly higher temperature cure (and temperature capability) than AF126. Finally,

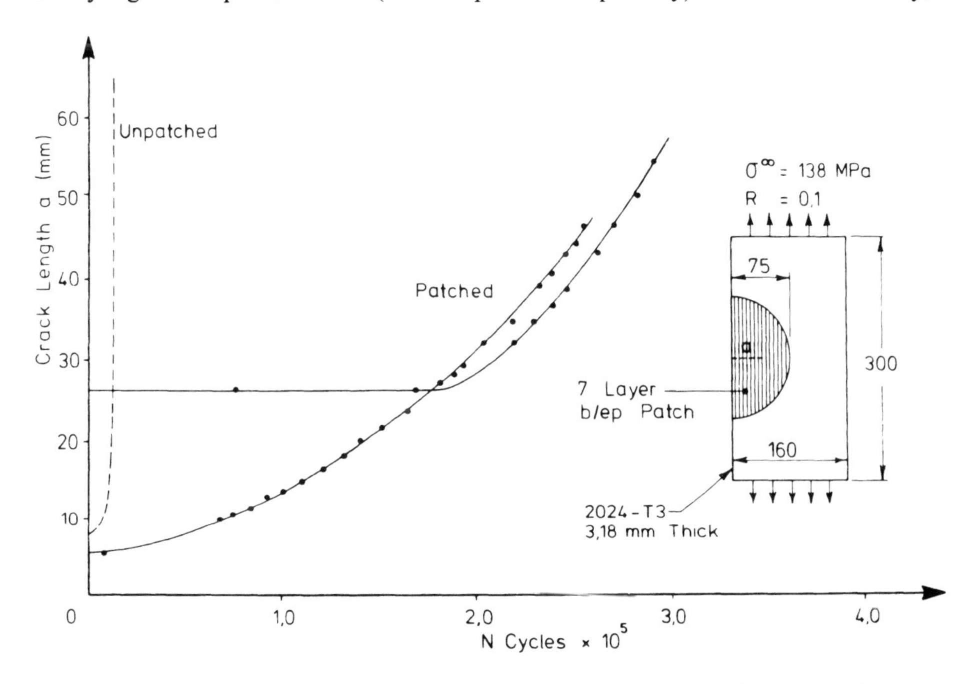

Figure 6.23. Plots of crack growth versus cycles for patched specimen (AF126 adhesive) solid lines and unpatched specimen dotted lines.

to study the influence of high bond temperature, (independent of the adhesive cure temperature) some AF126 specimens were subjected to further heat-treatment at 175°C.

In our fatigue test two similar specimens were simultaneously tested, joined together as a honeycomb sandwich panel. The aim of this configuration is firstly to minimise curvature, following patching due to the residual stress σ_T. Thus the patches were bonded at the same time as the panels were bonded to the honeycomb core. The second aim of the configuration is to minimise the bending of the panels which would otherwise occur during testing; the moments which cause the bending arise from the displacement of the neutral axis of the metal panels by the patch. The level of support provided by the honeycomb configuration against bending due to the residual-stress and the development of bending moments is considered to be a reasonable simulation of the support that would be provided by typical military aircraft structure. Agreement in fatigue performance of the two specimens was generally good. Crack growth in the patched panels was monitored through the patch using an eddy current procedure and following testing, the panels were subjected to X-ray radiography (using an X-ray absorbent fluid, tetrabromoethane) to detect disbonding or delamination of the patch. In many panels no evidence for significant disbonding or delamination was found – some, however, delaminated locally in the b/ep ply close to the crack.

Some of the findings and conclusions of this study are as follows:

(i) Patching efficiency is very high; Figure 6.23 which compares crack growth data for typical patched and unpatched specimen tested under the same nominal loading conditions.

(ii) Significant retardation in crack growth is experienced with the longer (25 mm) starting crack, but none with the shorter 5 mm starting crack (Figure 6.23). This difference in behaviour between the two crack sizes (borne out by practical experience described later) can be explained as follows. The degree of retardation of crack-growth following patching depends on the reduction in stress intensity range from ΔK_a to ΔK_R. Since the minimum stress intensity range ΔK_R is expected to be independent of crack size, the reduction in stress intensity range for the large crack and hence the degree of retardation, is much greater than for the small crack.

(iii) Crack growth (da/dN) under the patch is parabolic in some cases – see for example the short crack in Figure 6.23. This implies (on the basis of equation 6.26) that ΔK_R for the patched specimen is not totally independent of crack length a. However, the dependency of ΔK_R on a is not great so the general argument given above is correct to first order.

(iv) The degree of retardation is significantly reduced when relatively high temperature treatments are used. This behaviour evident in Figure 6.24, is considered further in the next section.

(v) Use of an adhesive curing at the highest temperature (175°C) leads to a slightly increased rate of crack growth. This is possibly due to the higher level of σ_T when the patch is applied at 175°C, since further reference to Figure 6.24 shows that when the patch is applied at 120°C followed by a heat-treatment at 175°C no increase in growth rate is apparent – even though the degree of retardation is

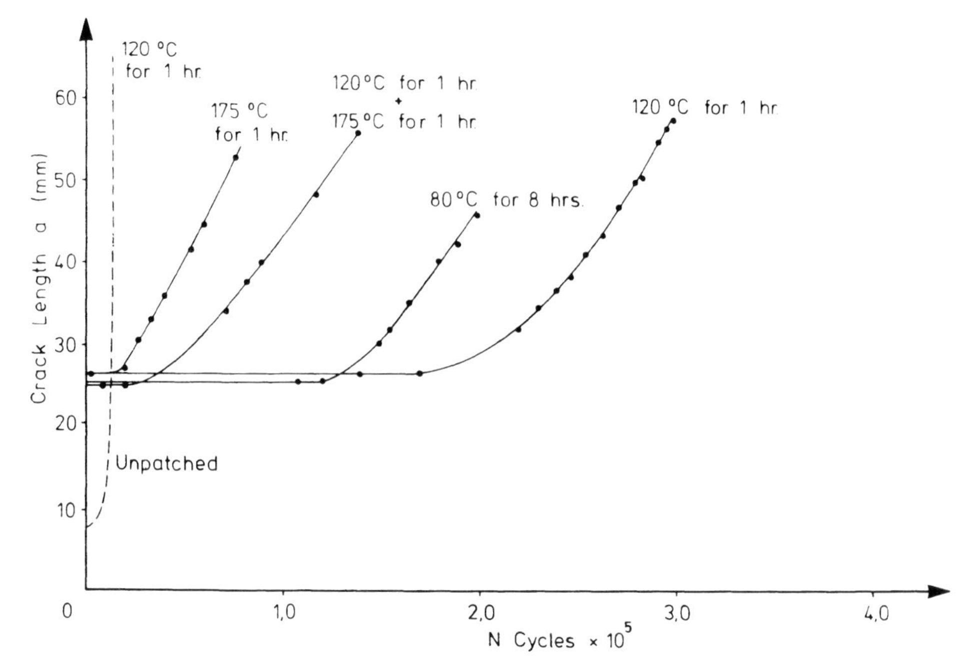

Figure 6.24. Plots of crack growth versus cycles for patched specimen subjected to various heat-treatment conditions as indicated during patch application.

similar. The theoretical value of σ_T following patching with AF126 adhesive is approximately 67 MPa (case 1, Table 6.11) whereas with FM300 σ_T is approximately 102 MPa.

In order to provide an approximate quantitative assessment of ΔK_R for comparison with ΔK_∞, the relationship between ΔK_a and $\mathrm{d}a/\mathrm{d}N$ was experimentally determined for the unpatched specimen. Data for three ratios of minimum to maximum stress (R values) are plotted in Figure 6.25. Using crack growth data for the short crack (Figure 6.23) the expected stress intensity ΔK_R was then read off from Figure 6.25 for the two extreme R values 0.1 and 0.6 and plotted in Figure 6.26. As shown below, these R values bracket the effective R value in the patched specimen R_R.

Firstly, for the unpatched panel tested under the same nominal stress conditions as the patched panel we have that

$$R = \sigma^\infty_{\min}/\sigma^\infty = 13.8\,\text{MPa}/138\,\text{MPa} = 0.1,$$

while, for the patched panel, the maximum stress under the patch is $\phi\sigma^\infty$ and the minimum stress is $\phi\sigma^\infty_{\min}$ (see section 6.4) so that, allowing for the residual-stress σ_T

$$\begin{aligned} R_R &= (R\phi\sigma^\infty + \sigma_T)/(\phi\sigma^\infty + \sigma_T) \\ &= (10.4\,\text{MPa} + 67\,\text{MPa})/(104\,\text{MPa} + 67\,\text{MPa}) = 0.45. \end{aligned}$$

The value of σ_T is obtained from Table 6.10, case 1 and ϕ is obtained using equation 6.14b for the conditions listed in case 1. Figure 6.26 shows the calculated

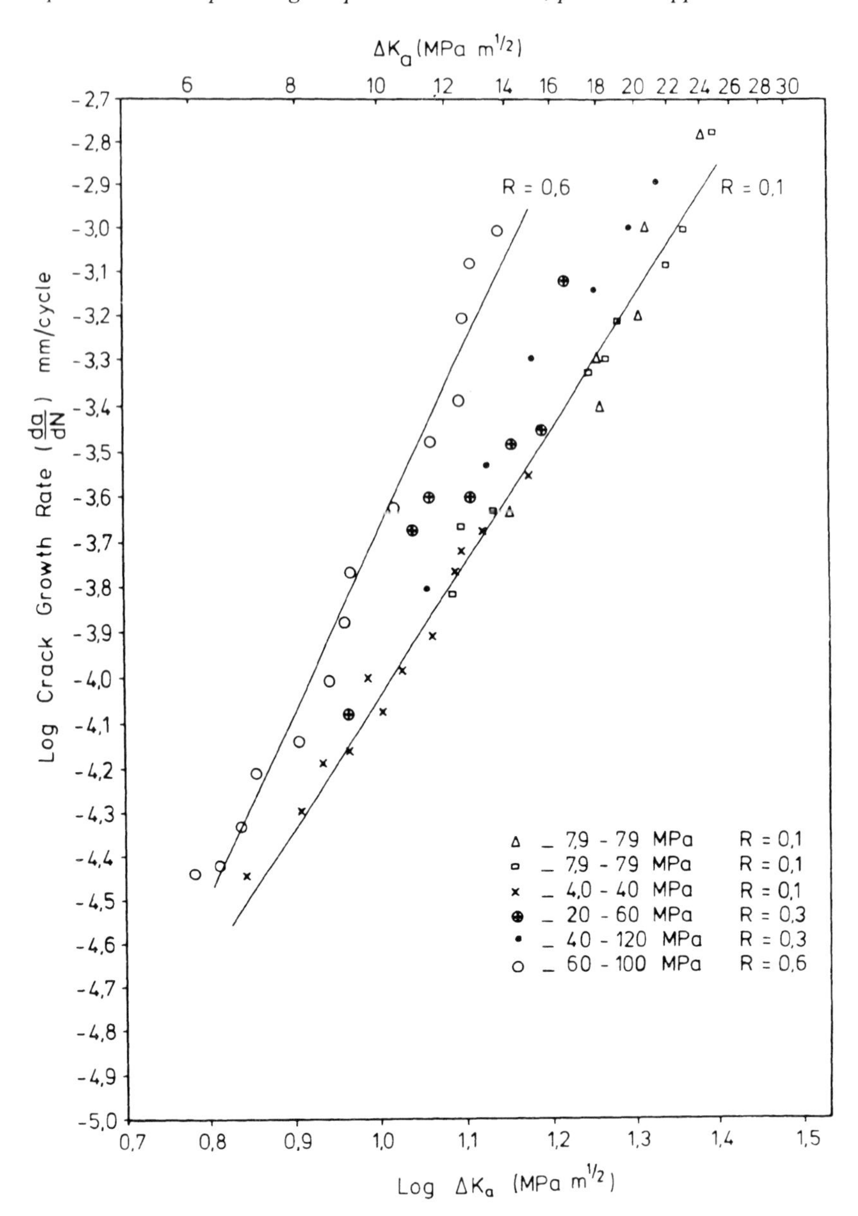

Figure 6.25. Plots log crack growth rate (da/dN) versus log ΔK_a for R values as indicated.

curve for ΔK_a, versus a, the estimated minimum and maximum values for ΔK_R (with or without residual-stress) and the theoretical values for ΔK_∞ and $\Delta K'_\infty$ taken from Table 6.10 case 1.

Although no claim is made that the method used here to estimate ΔK_R is other than approximate (particularly since other factors could influence crack growth – such as environmental isolation of the crack by the patch) the following conclusions are reached:

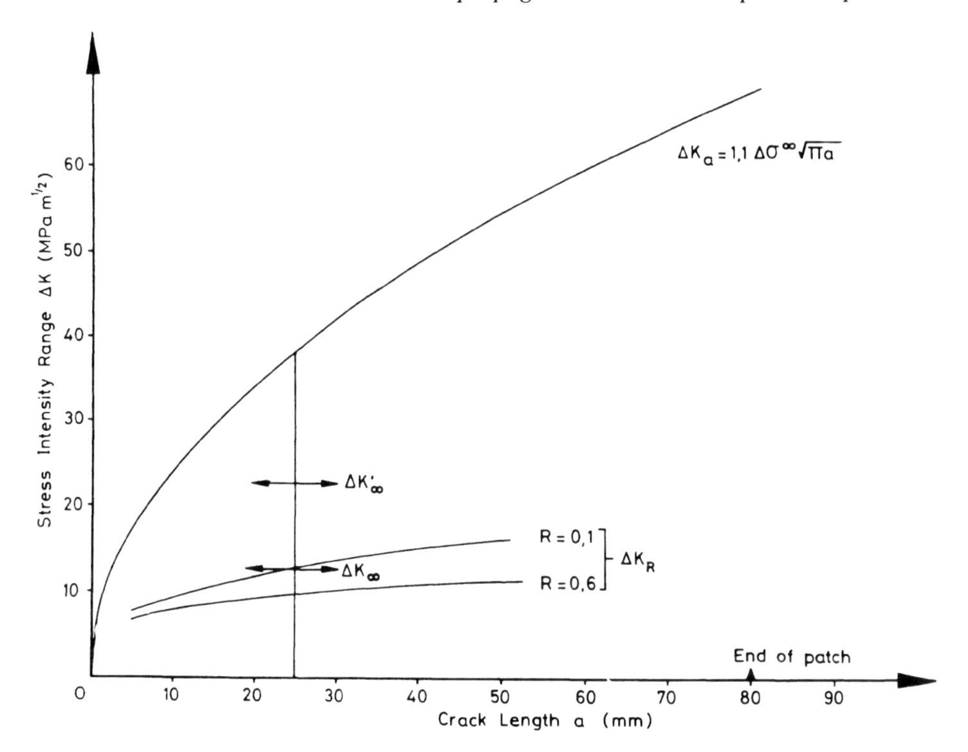

Figure 6.26. Plots of apparent stress-intensity range ΔK_R (derived from Figure 6.25) for the specimen with the 5 mm starting crack (plotted in Figure 6.24) versus crack length a. Also shown is the calculated relationship between ΔK_a and a and the theoretical values for ΔK_∞ and $\Delta K'_\infty$, taken from Table 6.10.

(i) Agreement between ΔK_∞ and ΔK_R is sufficiently good for ΔK_∞ to be usefully used for design estimations. The value of ΔK_∞ ($\sim$ 12.5 MPa m$^{1/2}$) is only in fair agreement with that obtained from the 3D finite-element analysis for this specimen configuration, Chapter 4 – viz. 7.5 MPa m$^{1/2}$.

(ii) The value of $\Delta K'_\infty$ which allows for plasticity in the adhesive clearly overestimates ΔK_R.

(iii) The influence of σ_T on ΔK_R is not great, about 10%, and may not be a concern in patching applications, particularly since σ_T would be lower in most practical situations – e.g. see Table 6.11).

(iv) ΔK_R varies (to a small extent) with crack size a in contrast to theoretical expectation.

6.6.2 Influence of adhesive cure or heat-treatment temperatures on unpatched specimens

Further tests were performed on edge-notched specimens to assess the influence of heat-treatment, under simulated bonding conditions, on retardation behaviour. These experiments involved the following steps:

(a) Specimens were cyclically stressed at 138 MPa until the crack length was 25 mm, as in the patching experiments.

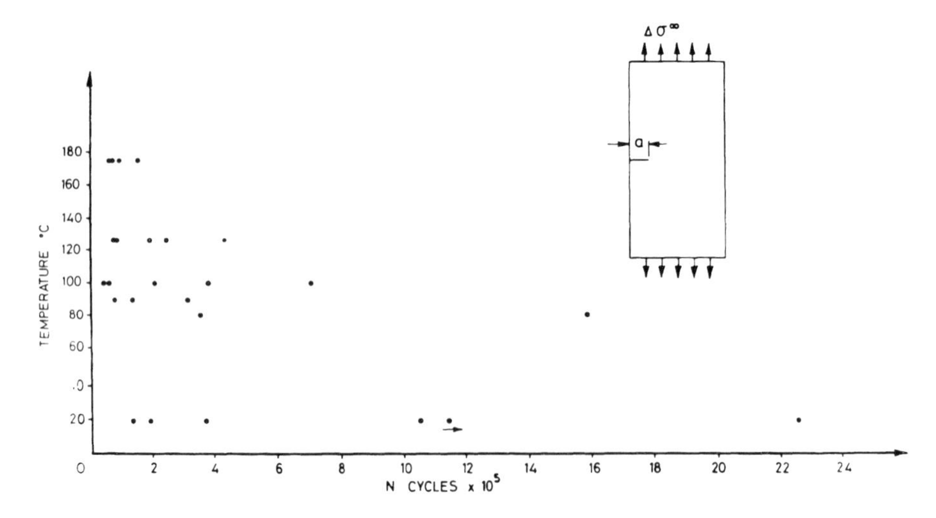

Figure 6.27. Plots of life after heat-treatment of unpatched edge-notched panels, initially cracked by cyclic loading to 25 mm at 138 MPa, then, following heat treatment at the temperatures indicated, retested at 79 MPa.

(b) The specimens were then heat-treated at one of the several bonding temperatures under investigation (including no heat-treatment, simulating an adhesive curing at ambient temperature).

(c) Finally, they were returned to the fatigue machine and cyclically stressed at 79 MPa ($R = 0.1$); the lower stress level simulates the reduced ΔK, that may have been achieved by patching.

Figure 6.27 plots the life of each specimen against the temperature of its heat-treatment condition. Since the remaining life, once crack initiation occurred, was only of the order of 2×10^3 cycles, Figure 6.27, effectively, plots lives to reinitiation of crack-growth or the degree of retardation.

From the results of the above experiment, the following observations are made.

(i) The scatter in the retardation observed greatly increases as the temperature of heat-treatment decreases below 175°C.

(ii) The minimum retardation is approximately the same at all levels.

(iii) The maximum retardation shows a fairly dramatic reduction above 90°C and is smallest after the highest heat-treatment temperature of 175°C.

The observations concerning the influence of temperature on the reduction in retardation in the patched specimens and the reduction in maximum retardation in the unpatched specimens can be explained in terms of the plastic zone effects. It appears likely that the heat-treatment relaxes the level of the elastic constraint, probably by a process of creep recovery in the plastic zone.

The scatter of the results noted in the unpatched specimen (and to a much lesser extent in the patched specimen) is difficult to explain. One possibility is that, the high ΔK employed on the initial testing induces multiple cracking, either on parallel planes or ahead of the crack tip, in some of the specimens, thereby lessening or negating the influence of the plastic zone.

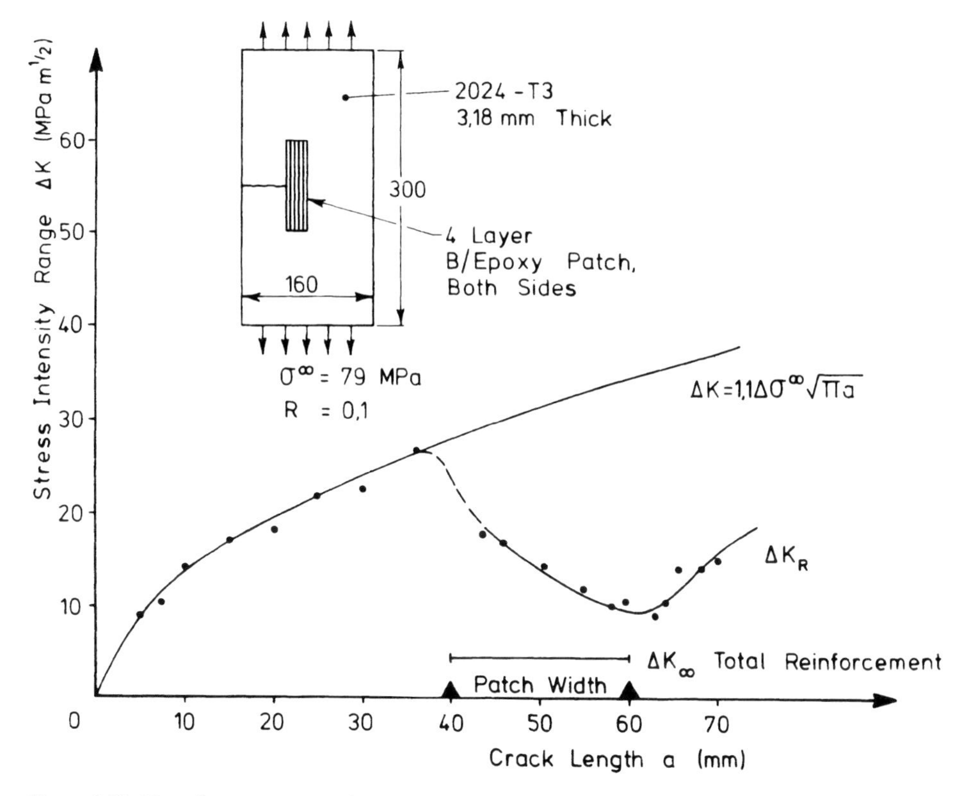

Figure 6.28. Plot of apparent stress intensity range (ΔK_R) versus crack-length (a) for an edge-notched specimen with tip reinforcement.

It is clear that, at least for 2024T3, adhesives which can be cured below 100°C are desirable if benefit is to be obtained from retardation.

6.6.3 *Patch specimen – tip reinforcement*

Specimens with the configuration depicted inset in Figure 6.28 were tested at constant stress amplitude to assess patching behaviour with tip reinforcement. In this case, patches were applied with AF126 prior to crack growth in order to avoid complications due to retardation. The crack growth data were analysed to obtain ΔK_R versus a (as described for the specimens with total reinforcement) and the results are plotted in Figure 6.28.

It can be seen that, prior to the crack reaching the patch the values of ΔK_R are in good agreement with ΔK_a, which confirms, that the patch has little far-field influence on crack growth. Once the crack has reached the patch the minimum value of ΔK_R occurs at the far edge of the patch. It appears as though the value of DK_R would asymptote to ΔK_∞, if the patch were wide enough – in this case if the patch width was increased from 20 mm to 40 mm. These results are considered further in Chapter 5 where a detailed analysis is developed.

Table 6.12. Some applications of crack-patching.

Cracking	Material/ thickness mm	Component	Aircraft	Comments
Stress-corrosion	7075-T6/2.6	Wing plank	Hercules	Over 400 repairs since 1975, no crack growth.
Fatigue	Mg alloy/10 MSR	Landing wheel	Macchi	At least one lifetime extension. Bond durability is main problem.
Fatigue	AU4G1/1 (French alloy similar to 2024T3)	Fin skin	Mirage	Satisfactory service since 1978.
Fatigue	AU4SG/3.5 (French alloy similar to 2014T6)	Lower wing skin/ fuel decant hole	Mirage	Over 150 repairs since 1979.
Fatigue	AU4SG/3.5	Lower wing skin/ fairing attachment hole	Mirage	Recent repair.
Stress-corrosion	7075-T6/19	Console-truss	F111	Satisfactory service since 1980.
Fatigue	2024-T3/0.6	Upper wing skin	Nomad fatigue test article	Over 105900 simulated flying hours, no growth.
Fatigue	2024-T3/0.6	Door frame	Nomad fatigue test article	Over 106619 simulated flying hours, no growth

6.7 Applications of crack-patching

Some of the crack-patching repairs developed by ARL for RAAF military aircraft are listed in Table 6.12. Only the applications to stress-corrosion cracks in Hercules wing planks and fatigue cracks in Mirage wing skins are described here since these examples (the first quite simple and the other considerably more involved) are sufficient to demonstrate the benefits of the repair technology to aircraft structural components. It should be mentioned that, in most cases, practical exploitation occurred well prior to completion of the background studies described here. However, while many improvements could have been made with hindsight, the basic approach would not have differed significantly. Future major improvements will almost certainly be based on improved adhesive technology, surface-treatment and patch application procedures.

6.7.1 Hercules aircraft wing planks

The wing planks are machined from solid 7075 T6 rolled plate material to produce wing skins with integral reinforcing risers, about 2.6 mm thick. Internal wing structures, at a separation of about 450 mm are riveted to the risers. Stress-corrosion cracks frequently initiate in the risers from rivet holes (possibly due to rivet pressure inside the hole) and propagate in the risers, parallel to the wing surface, as illustrated schematically in Figure 6.29. Because of the huge cost and time involved in disassembly of the wing it is most desirable that the cracks are repaired *in situ*. Fortunately there is just enough working space inside the wing for this operation to be feasible.

The standard procedure when cracks reach a length of about 200 mm is to rivet a reinforcement bracket onto the riser, Figure 6.30. This approach has the disadvantage that it causes damage due to the additional fastener holes (which can

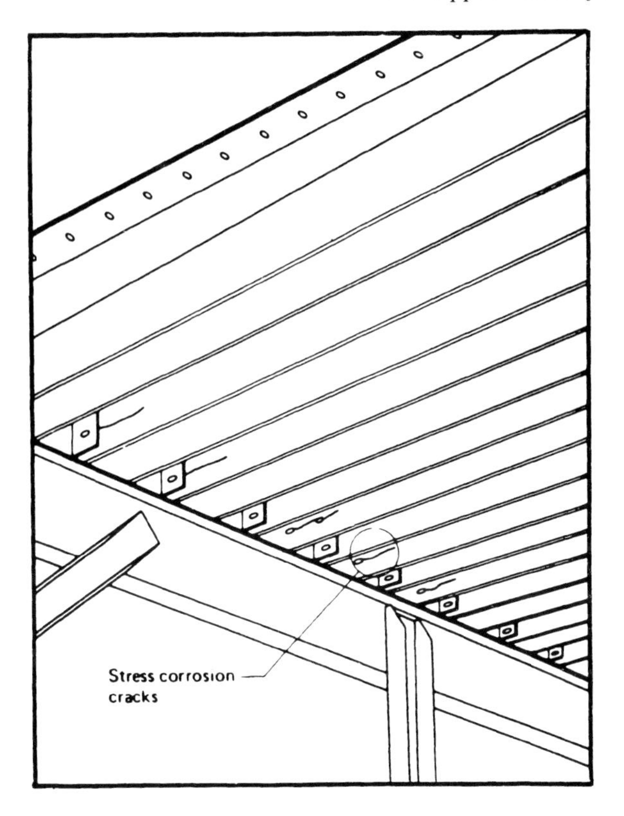

Figure 6.29. Diagram of underside of Hercules upper wing plank showing the position of stress-corrosion cracks.

encourage further cracking), results in local-stiffening of the riser, makes the cracks difficult to detect and monitor and is quite time-consuming.

By contrast to the above, the crack-patching procedure [2] is simply to bond a b/ep patch (4 plies thick) adhesively onto each side of the riser with the fibres spanning the crack, using adhesive AF126, Figure 6.31. In view of the unknown level of residual-stress involved in driving the initial stress-corrosion cracks it was not possible to undertake a patch design study for this repair. It was assumed that initiation and growth of the cracks resulted from a wedging action by the rivets resulting in transverse tensile stresses, and simple experiments, as illustrated in Figure 6.32 were employed to evaluate the effectiveness of patching. These experiments established that the reinforcement of the crack and the sealing action of the adhesive both contributed to the effectiveness of the patch in arresting stress-corrosion cracks.

Alumina grit-blasting was chosen as the method for surface-treating the riser prior to adhesive bonding the patch; the more advanced surface-treatments or the use of silane primers had not yet been developed. A vacuum attachment for the grit-blaster was developed to remove the abrasive and avoid excessive contamination of the wing.

Heat to cure the AF126 adhesive was supplied by silicone rubber heating tapes;

Figure 6.30. View of underside of Hercules upper wing plank, showing example of standard repair scheme based on rivetted aluminium alloy repair patch, applied to a 250 mm stress-corrosion crack.

these also act as compressible layers to help maintain pressure during curing of the adhesive. Pressure was applied to the patch through the heating tapes by hand-closed toggle clamps.

The above repairs are now made on a routine basis by RAAF technical personnel. Typically, a bonded repair takes just one man day compared to a nominal six man days for the standard procedure. To date, there have been no observations of bond failure or crack growth following repair; many repairs date back to 1975.

6.7.2 Mirage wing skins

A severe fatigue-cracking problem developed in the lower wing skins in several Australian Mirage aircraft. The region of cracking is illustrated in Figure 6.33 which shows that it occurs from a fuel decant hole close to the intersection of the main spar and the root rib. The cracking appears to be related to a severe stress concentration associated with the proximity of a rivet hole (used to anchor an attachment nut) to the fuel decant hole. Since the wing skin forms part of the main torsion box, it is in a state of shear which results in cracks propagating normally to the principal tensile stress at an angle of 45° to the spar as indicated in Figure 6.34. The skins consist of aluminium alloy AU4SG (a French alloy, similar to 2014 T6) about 3.5 mm thick with integral reinforcing risers. More recently an adjacent fairing attachment hole region was also found to be prone to cracking. However, the problem in this area is substantially less critical than that associated with the fuel decant hole.

Figure 6.31. View of underside of Hercules upper wing plank, showing b/ep repairs bonded over several stress-corrosion cracks, about 50 mm long.

A b/ep crack-patching repair was proposed for repair of both problems. The advantages over the conventional alternative of a bolted aluminium plate repair were given as the following: (i) the b/ep repair (due to its highly directional properties) should introduce no significant local elevation of strain in the loaded spar, (ii) mechanical damage would not be inflicted on the skin (no fastener holes), (iii) the thickness of the skin (including the repair) would impose no system or fastener clearance difficulties, (iv) eddy-current procedures could be employed to check for crack growth, and (v) implementation of the repair under field conditions (with fast turn-around time) should be possible.

In view of the high stresses involved and the significance of the repair, it was decided to proceed initially with a feasibility study based mainly on design studies and detail testing. If feasible, the aim was to apply the repair to all aircraft, to repair cracked wings or to act as a preventative measure for uncracked wings. The final configuration of the repairs is shown in Figure 6.33.

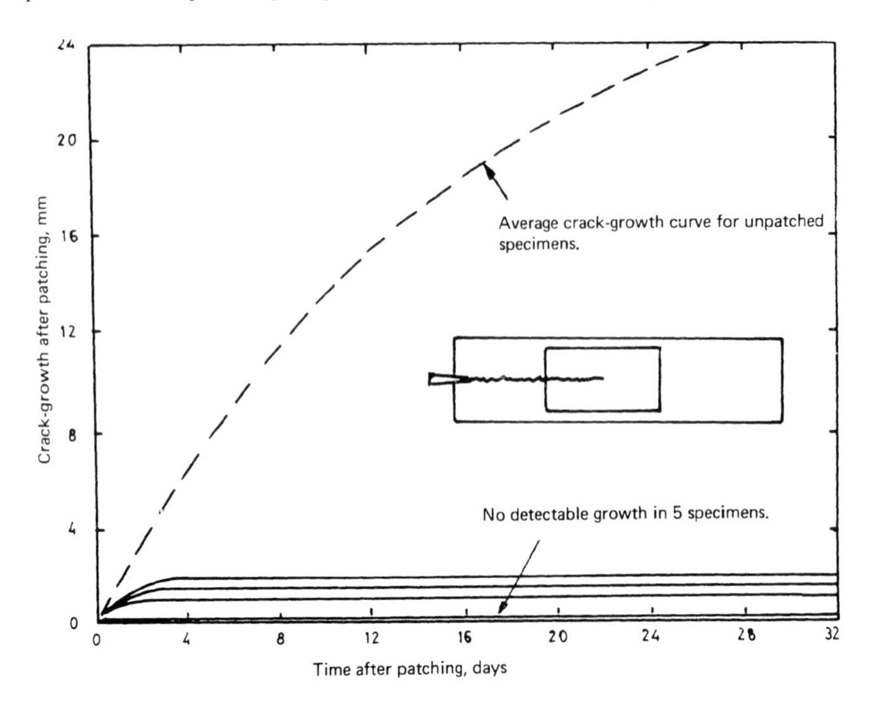

Figure 6.32. Experimental plots of crack growth versus time for wedge-loaded stress corrosion specimens (cut from the wing plank material).

The decant hole patch chosen comprised of a seven layer unidirectional laminate internally stepped, (i.e. with the largest layer on the outside) around its periphery. The aim of this construction is to reduce interlaminar shear stresses and provide good external smoothness. The fairing hole patch was similar in configuration but only six layers thick. The b/ep patches were to be bonded with AF126, employing phosphoric acid gel-anodising as the method for surface-treating the wing skin.

The finite-element procedure was employed for the design studies, as described in Chapter 4. The conclusions of this study were that the strains in the patch and (apart from a few small regions) in the adhesive were acceptably low and that the reductions in stress-intensity for the cracks was very significant.

Repair qualification

Two main tests were performed to qualify the repair, (i) a strain survey on a Mirage wing with a decant hole repair patch installed to check that the patch did not significantly elevate the local strain in the spar, and (ii) fatigue tests on aluminium alloy panels configured to simulate the cracked areas to check that (a) the predicted reduction in ΔK could be achieved, and that the patch/adhesive system could endure the fatigue loading. The strain survey [31] confirmed the expectation that no significant strain elevation occurred in the spar under load after patching.

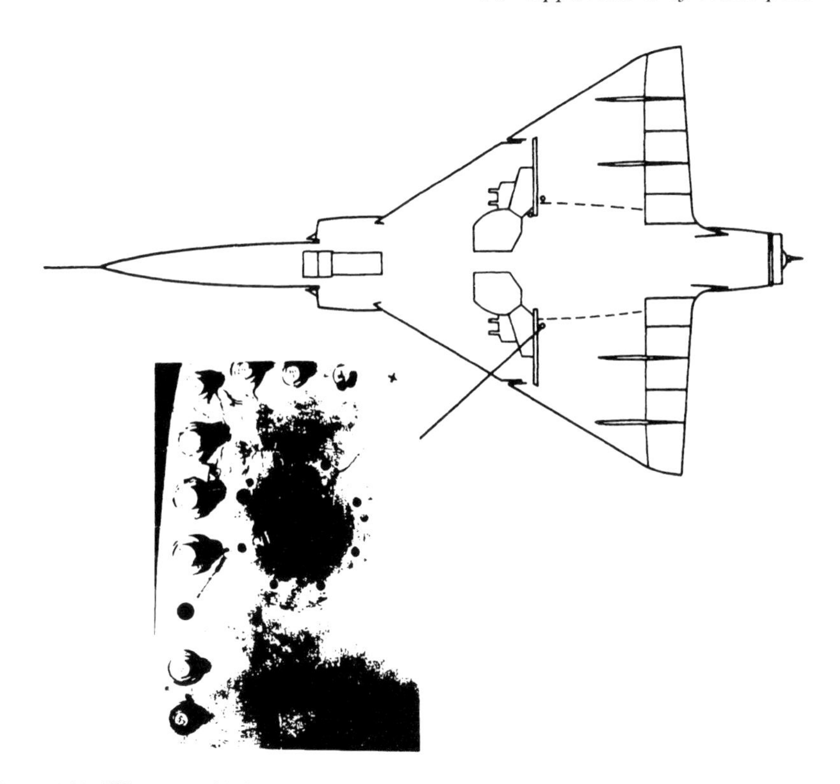

Figure 6.33. Silhouette of Mirage III aircraft showing where fatigue cracks develop in some aircraft. Inset: fuel decant hole region showing nature of fatigue cracking.

Table 6.13. Load and stress sequence employed for the patched panels simulating the Mirage decant hole region. Note sequence is symmetrical about load level 1.

Load Level	Max Nominal g Load	Min Nominal g Load	Max Stress MPa	Min Stress MPa	Number of Cycles
9	7.8	0	167	0	0.3
8	7.4	0	162	0	1
7	6.5	0	149	0	31
6	5.25	0	130	0	381
5	4.0	0.05	110	5	579
4	3.2	0.13	95	9.7	341
3	2.75	0.18	85	11.8	870
2	2.25	0.22	74	13	755
1	1.75	0.38	62	20.4	10,000
1	1.75	0.38	62	20.4	10,000
2	2.25	0.22	74	13	755
3	2.75	0.18	62	11.8	870
etc	etc	etc	etc	etc	etc

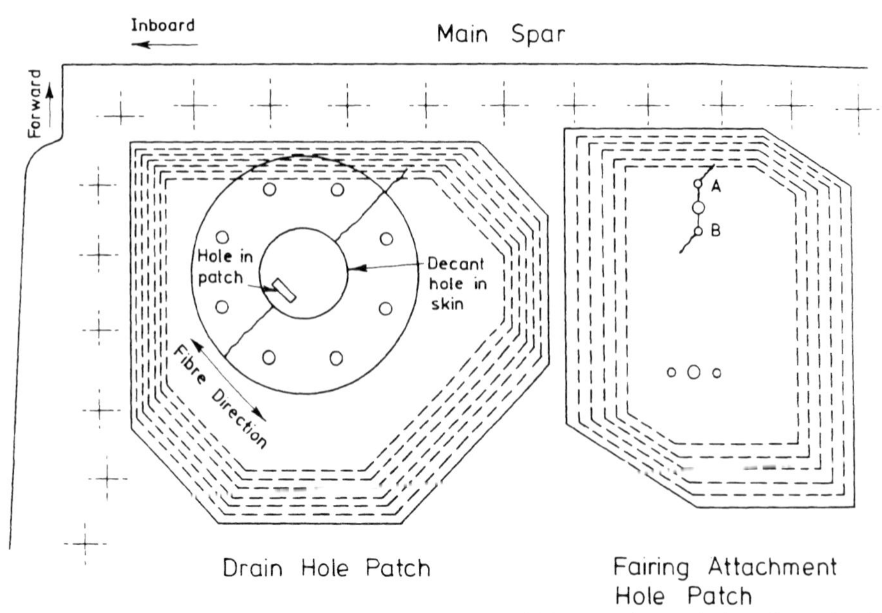

Figure 6.34. Schematic of (a) fuel decant hole, and (b) fairing attachment hole regions, showing cracking and outline of b/ep patches.

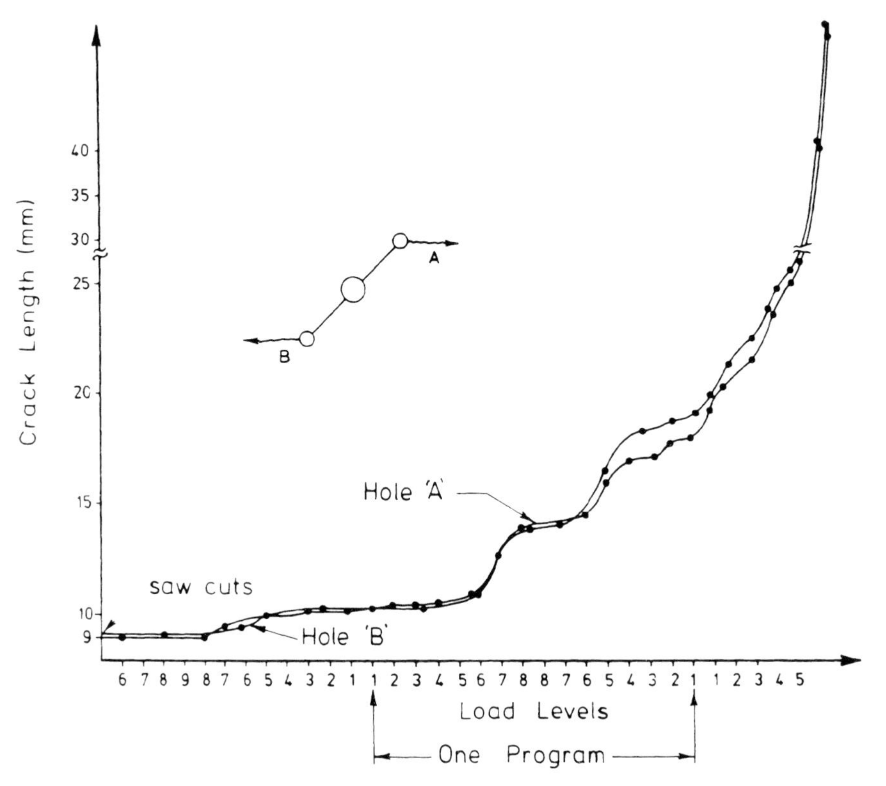

Figure 6.35. Plot of crack length versus cycles for an unpatched specimen with the cracks initiating from the fairing hole region.

Tension-tension fatigue tests [4] [32] were undertaken on panels made from alloy 2024T3 and later, from BSL-104 (similar to AU4SG). Only selected results on BSL-104 panels are described here, since these most closely represented the repair situation. The tests were undertaken using the block loading sequence listed in Table 6.13. Each complete sequence nominally corresponds to a year's aircraft operation. This loading sequence was constructed from Mirage usage data described in reference [32]. Only uniaxial tension fatigue loads were applied which were intended to represent the principal tensile stress component of the shear stress present in the wing (compression stresses at 90° were thus neglected).

Cracks were initiated from saw-cuts and propagated by fatigue under the block loading sequence to the desired size. The usual size was up to 30 mm from each side of the decant hole and up to 15 mm from each hole for the fairing hole region. During application of the patches, the panels were restrained at their ends to simulate the restraint to thermal expansion expected in the wing and, during testing, the panels were restrained against secondary bending by using a vacuum box technique [4]. For comparison with the results on cracked and patched panels, tests were initially performed on some unpatched panels. Figure 6.35 plots the results of crack growth for cracks initiated from a region of a panel with the fairing hole configuration. This shows that the panel could not withstand more than one block-loading sequence before crack growth became catastrophic.

Results of crack size versus programs of loading are plotted in Figure 6.36; growth under the patch was followed using an eddy-current procedure. In the test

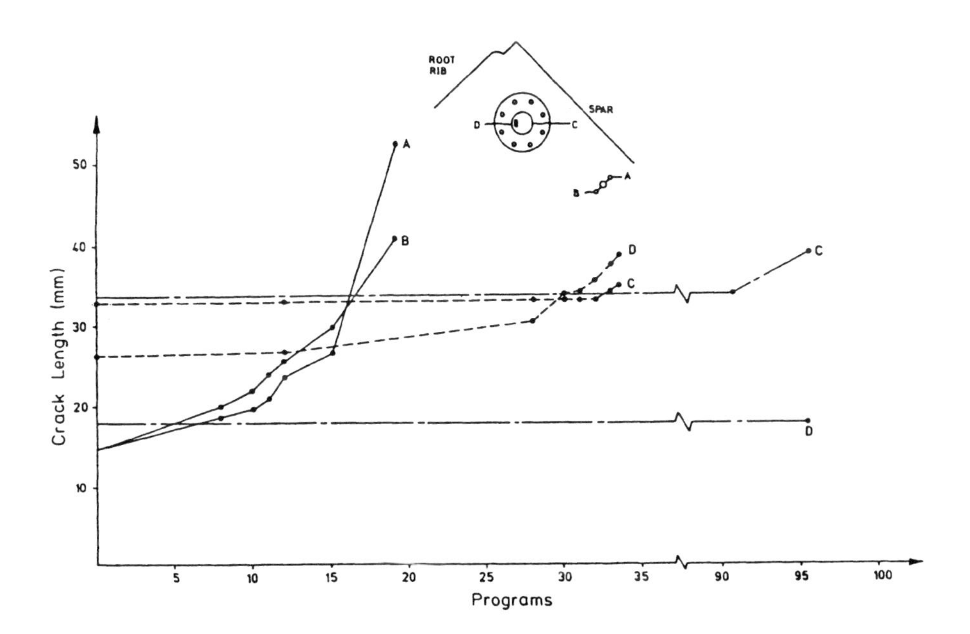

Figure 6.36. Plots of crack growth versus cycles for patched panels. Solid lines are results from one panel and dotted lines from another.

represented by the dotted lines, b/ep patch repairs were made only to a simulated decant hole region, while in the test represented by the solid lines, repairs were made to simulated fairing hole and decant hole regions.

Comparison of the above results with those for the unrepaired panel, Figure 6.35 shows that a major improvement in crack propagation performance is obtained in all cases. Comparison of the various results in Figure 6.36 shows that they differ in the degree of retardation of crack growth. In particular the short (simulated fairing) cracks shows very little retardation. The result is in accord with that reported for the comparison between long and short cracks in Section 6.6 and can be explained by the reasons given there – even though the tests reported in this section were conducted under programmed loading.

On the basis of the above, and other tests, it was decided to proceed with implementation of the repair program to RAAF aircraft.

Repair implementation-decant hole repair

Patches were manufactured by a modified autoclave moulding procedure from pre-preg tape. Because of the large number of patches required, a numerically controlled laser was first employed to cut the tape into plies containing the design holes and cut-outs. The bonding surface of the patches was prepared both by removal of peel ply and by alumina grit-blasting and covered with a film of

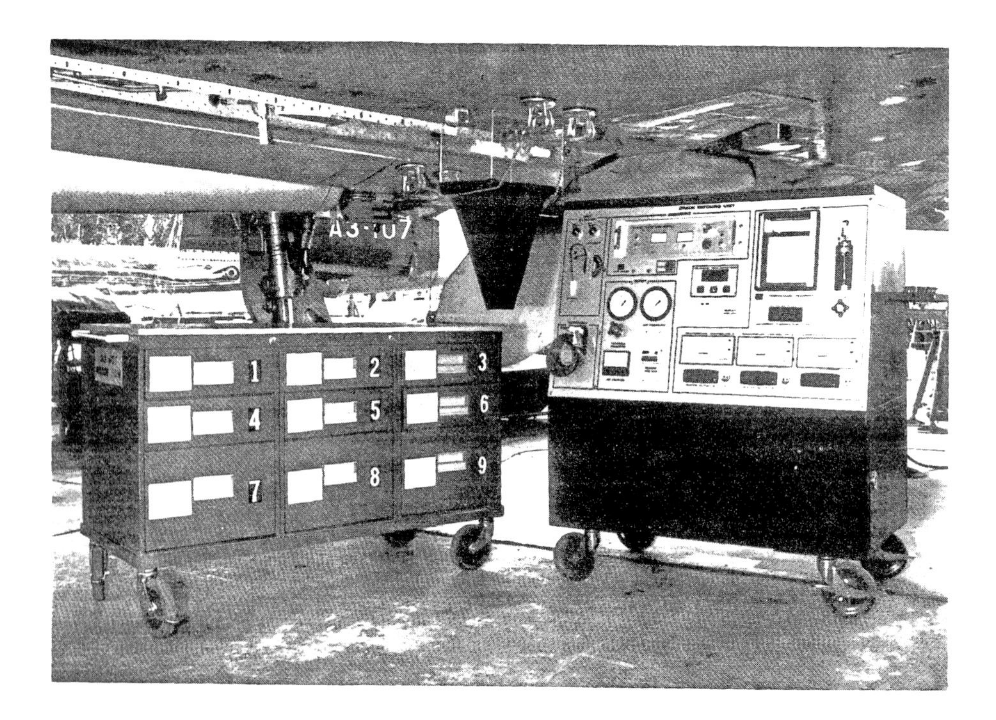

Figure 6.37. Photograph of crack patching unit developed for application of b/ep patches to Mirage wing skins.

(uncured AF126) for shipment, under refrigeration to the RAAF bases where the repairs were applied.

To aid implementation of the repair by (ARL Trained) RAAF technical personnel, a crack-patching ground support unit was developed, Figure 6.37. This unit contains all necessary equipment including (i) a grit-blaster, (ii) an anodising unit, (iii) temperature controllers to control cure temperature at $\sim 120°C$, and (iv) a patch pressurisation system based on a hydraulic jack.

There are eight major steps in the repair procedure [33]:

(a) Aircraft preparation: Seal fuel leaks and detail the crack positions.

(b) Blanking Off: Seal the decant hole with an aluminium disk bonded with polysulphide rubber.

(c) Preheat Drying: Dry out any fuel which has not been removed. During this operation the wing tanks are purged with nitrogen to eliminate risk of ignition of fuel vapour.

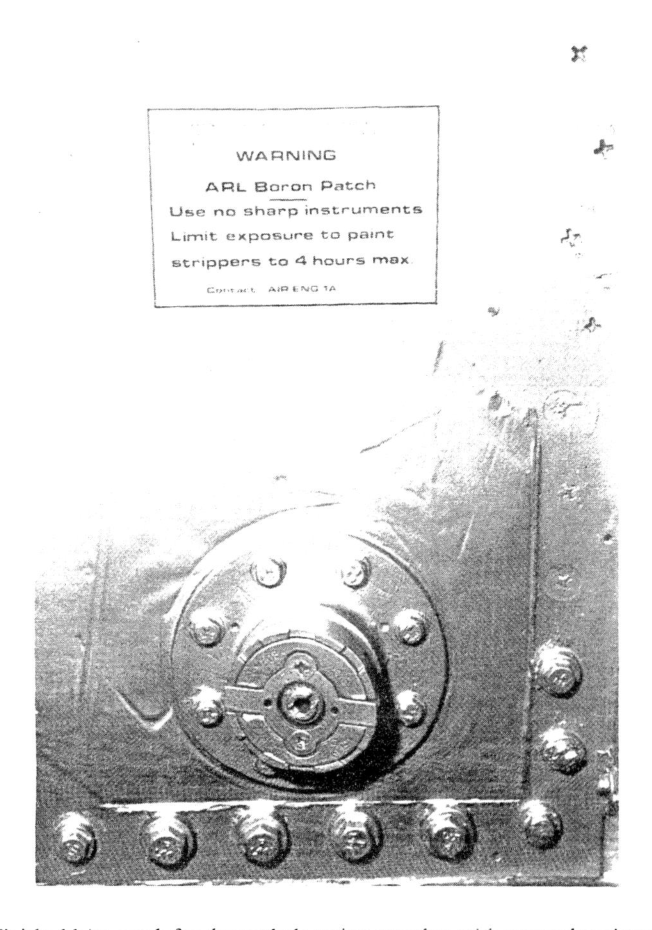

Figure 6.38. Finished b/ep patch for decant hole region complete with external environmental protection in position on a Mirage wing.

(d) Masking Off. Cover regions surrounding the repair with adhesive tape to prevent contamination with phosphoric acid gel.

(e) Grit Blasting: Surface is first cleaned with detergent followed by solvent until water break clean, then thoroughly grit blast with fresh alumina.

(f) Anodise the Surface: Use the phosphoric acid gel anodising procedure-check for anodise condition, repeat if required.

(g) Adhesive Curing: Bond patch to wing at about 110°C with direct pressure. Again nitrogen flushing is employed to eliminate fire risk.

(h) Final Sealing: Seal various regions and cover patch with a protective layer of aluminium to act as a moisture barrier to minimise environmental degradation, particularly for protection against paint stripper.

The finished repair is shown in Figure 6.37. Generally the repairs were applied during routine service of the aircraft so as to maximise aircraft availability, and required two to three working days. The patching program commenced with experimental trials in December 1979 and was completed by October 1981 with patches applied to about 180 Mirage wings. To date, the repair appears to have been highly successful, although a few cases of crack growth following patching were reported for wings with large cracks, further growth has ceased in all but a few cases.

References

[1] Baker, A.A., *Composites* 9 (1), 11–16 (1978).

[2] Baker, A.A. and Hutchison, M.M., Aeronautical Research Laboratories, Australia, Technical memorandum 366, (1976).

[3] Baker, A.A., *Composite Structures* 2, 153–181 (1984).

[4] Baker, A.A., Callinan, R.J., Davis, M.J., Jones, R. and Williams, J.G., *Theoretical and Applied Fracture Mechanics* 2, 1–5 (1984).

[5] Ratwani, M.M., Labour, J.D. and Rosenweig, E., Proceedings of the 11th ICAF Symposium, Paper No. 1.6, (1981).

[6] Scott, R.F. and Huculak, P. *Canadian Aeronautics and Space Journal*, 28, (2), 122–134 (1982).

[7] Kemp, R.M.J., Murphy, D.J., Butt, R.I., Wilson, R.N. and Phillips, L.N. *Fatigue of Engineering Materials and Structures* 7, (4), 329–344 (1984).

[8] Baker, A.A. *SAMPE Journal* 15, (2), 10–17, (1979).

[9] Davidson, R.G., Ennis, B.C., Mestan, F.C. and Morris, C.E.M., Materials Research Laboratories, Australia, Technical Report, OCD, 84/3 (1984).

[10] Parker, B.M. and Waghorne, R.M. *Composites* 13, 280–287 (1982).

[11] Hill, T.G., Unpublished Work, Aeronautical Research Laboratories, Australia.

[12] Lock, M.C., Horton, R.E. and McCarty, J.E., AFML-TR-78-104 (1978).

[13] Technical Brochure Silane Coupling Agents Dow Corning (1981).

[14] Stringer Motor Vehicle Engineering Establishment, U.K., Report 81506 (1981).

[15] Broughton, W. and Baker, A.A., Work to be published, Aeronautical Research Laboratories, Australia.

[16] Stone, M.H. and Peet T. Royal Aircraft Establishment, UK, Technical Memo, Mat 349 (1980).

[17] Baker, A.A., Davis, M.J. and Hawkes, G.A., Proceedings on the 10th International Committee on Aeronautical Fatigue (ICAF) Symposium, paper 4.3 (1979).

[18] Baker, A.A., Hawkes, G.A. and Lumley, E.J., Proceedings of the Second International Conference on Composite Materials, Toronto, Metallurgical Society of AIME (1978).

[19] Williams, J.G. *Stress Analysis of Polymers* Longmans, London, 225–243 (1973).

[20] Rose, L.R.F. *International Journal of Fracture* 18, 135–144 (1982).
[21] Jones, R. and Callinan, R.J., *Journal of Structural Mechanics* 8, (2), 143–149, (1980).
[22] Rose, L.R.F., *Int. J. Solids Structures* 17, 827–838 (1981).
[23] Baker, A.A., Roberts, J.D. and Rose, L.R.F., Proceedings 28th National SAMPE Symposium, 627–639 (1983).
[24] Althoff, W., Adhesion 5, 15–27 (1980).
[25] Jones, R. and Callinan, R.J. *Journal of Structural Mechanics* 8, 143–149 (1980).
[26] Jones, R. and Paul, J., Aeronautical Research Laboratories, Australia, Structures Report 402 (1984).
[27] Hertzberg, R.W. and Manson, J.A., *Fatigue of E ...ering Plastics*, Academic press (1980).
[28] Hart-Smith, NASA, CR-112235 (1973).
[29] Baker, A.A. and Packer, M.E., Work to be Published Aeronautical Research Laboratory, Australia.
[30] Broek, D., *Elementary Engineering Fracture Mechanics* Sijthoff and Noordhoff (1978).
[31] Hoskin, B.C. and Baker, A.A. Aeronautical Research Laboratory, Australia Structures Technical Memorandum 321, (1980).
[32] Van-Blaricum, T.J. Aeronautical Research Laboratories, Australia, Structures Technical Memorandum 404 (1985).
[33] Davis, M.J. and Roberts, J.D., Aeronautical Research Laboratories, Australia, Materials Technical Memorandum 373 (1981).

R.E. TRABOCCO, T.M. DONNELLAN AND J.G. WILLIAMS

7

Repair of composite aircraft

7.1 Introduction

The use of composite materials in aircraft has been rapidly increasing. The high specific strength and stiffness of composites results in significant weight savings when these materials are substituted for aircraft alloys. The use in aircraft has been extensively reviewed [1–3]. Because the technology is relatively new, composites have only recently been introduced into structural applications. Advanced composite structure was first introduced on military aircraft when the reduced weight was desired for improved tactical performance. Commercial aircraft manufacturers are also beginning to use composites in structural applications where reduced structural weight leads to significant fuel savings. Many secondary structural components including flaps, fairings, and floor panels have been constructed from graphite and kevlar based composites. Figure 7.1 shows the use of composites in two recent aircraft [3, 4].

The types of composites used in structural applications in the aerospace industry are usually fiber reinforced thermoset resins. The fibers impart the strength and stiffness to the composite and the resin matrix acts as a load transference medium. The combination of resin and fiber used in any particular application will depend on the operating environment of the component. Many texts are available describing the use of composites [5]. The matrix resin is chosen on the basis of the maximum operating temperature of the component in service. Currently used composites usually have epoxy resin matrices. In some areas, the thermal stability of epoxies is insufficient and other polymers such as polyimides and bismaleimides are used [6]. The fibers used include glass, boron, graphite, and aramid. The type of fiber used will depend on the mechanical properties needed in the application. Glass fibers are inexpensive and possess high strength but are relatively dense. Graphite fibers are most commonly used in structural aircraft applications due to their high strength and modulus and moderate price. Aramid fibers can be used to improve fracture toughness but the application must be one in which the relatively poor compressive strength of the aramid fibers does not affect performance. Boron fibers are relatively expensive and difficult to use but they may be used in areas

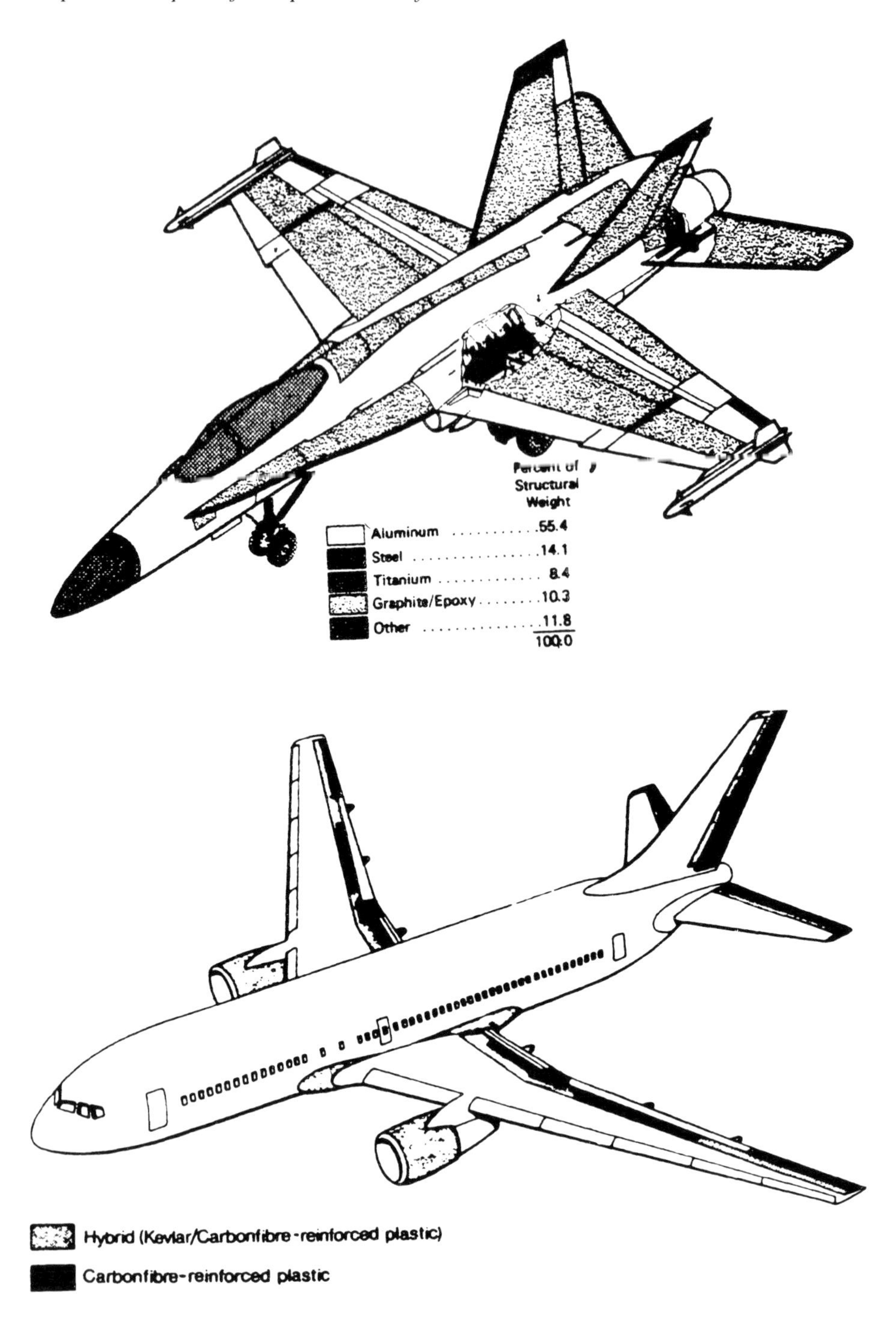

Figure 7.1. The use of composites in two recent military and commercial aircraft.

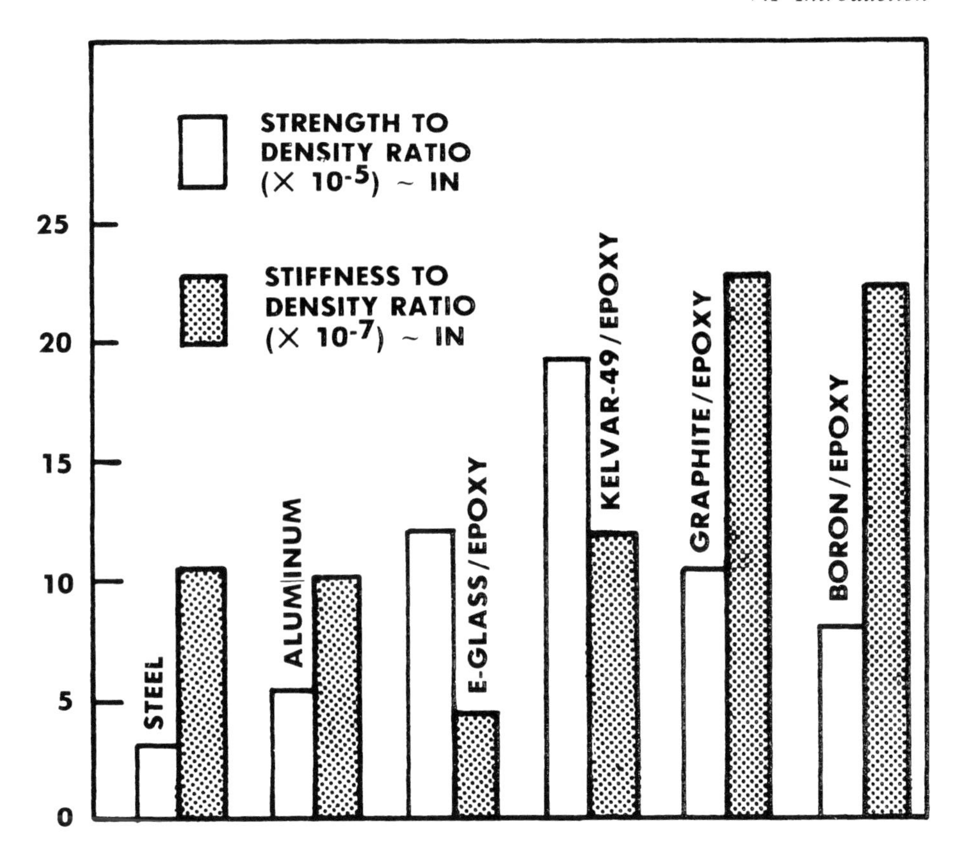

Figure 7.2. A comparison of specific strength and specific stiffness of composites and common metals.

where high compressive strength and stiffness are required. The relative properties of composites and common structural metal alloys are shown in Figure 7.2.

Design of composite components presents a marked departure from that of metallic materials [7–9]. Composite materials are essentially anisotropic nonhomogenous materials. The properties in a composite are developed by orienting the fibers in the load bearing directions. A composite can be custom designed to fit orthogonal loading conditions. They are not suited for out-of-plane loads, particularly those through the thickness. Generally, external loads are translated into lamina requirements by suitable transformation matrices; the lamina properties are then summed to determine the laminate requirements for the particular load situation. Some suitable criteria such as maximum stress or strain can be chosen for design limit criteria.

Composites behave differently than metals both under static and dynamic loading. Compression is more of a concern in composites because of the matrix contribution to this loading situation. Tensile failure is a manifestation of an accumulation of individual fiber failures. In tensile fatigue, the failure stress does not markedly decrease with the number of load cycles, while in compressive fatigue there is some damage accumulation which results in decreased stress capability with increased cyclic loading. Composites have little means of sustaining plastic deformation and disclose relatively small amounts of strain to failure compared to metals. Low-velocity, foreign-object damage can result in delamination of plies in a composite laminate with little evidence of such an occurrence on the impacted face. Composite anisotropy presents difficulties in the mathematical handling of fracture toughness, which is routine in metals. Some progress has been made in this regard with computer routines to handle the complex mathematics.

7.2 Composite fabrication

Composites are able to attain their exceptional properties because they represent a uniquely efficient combination of the component materials characteristics. Composites used in airframe applications are most commonly manufactured by laminating layers of pre-preg. A pre-preg consists of a set of reinforcing fibers which have been wetted or impregnated with uncured matrix resin material. During fabrication, layers of pre-preg material are stacked together in a particular sequence to attain strength in the specific directions required by the component design (see Figure 7.3 [10]). Composite processing involves consolidation of the pre-preg plies into a homogeneous material and reaction of the uncured matrix into a cross-linked structural polymer.

There are a number of techniques which can be used to process composites. These include autoclave, heating blanket, and press cure techniques. An autoclave is a pressure oven which also has the capability of drawing vacuum. Most composite components are fabricated in an autoclave with a suitable mold. Autoclaves typically are cylindrical chambers with electrical heating which is actively circulated for thermal uniformity. Part temperature is monitored with thermocouples placed adjacent to the component. Pressure is supplied to the chamber interior via air pressure or bottled gas. The composite is cured with a particular cure cycle, a combination of temperature and pressure required for that system. State-of-the-art epoxy materials are typically cured at 177°C for at least 1 hour in concert with 700 KPa of pressure. An example of a cure cycle used for processing a standard gr/ep system system is shown in Figure 7.4.

Cure of composites can also be achieved using a heating blanket and vacuum as a source of compaction. Such a procedure is used where large amounts of positive pressure are not necessary for compaction. Elevated temperature is still required to achieve adequate cross linking in the cured polymer. As with the previously described method, thermocouples are placed in the bagged laminate to control temperature and to avoid overheating. Such an approach would be necessary if an autoclave were not available.

Continuous reinforced composites can also be fabricated in elevated temperature presses. Heating elements in the platens supply the thermal energy for cure. Pressure is supplied directly through the platens to achieve the desired degree of compaction in the laminate. Mold forms can be used in conjunction with such presses to achieve a particular configuration. Vacuum can be incorporated into

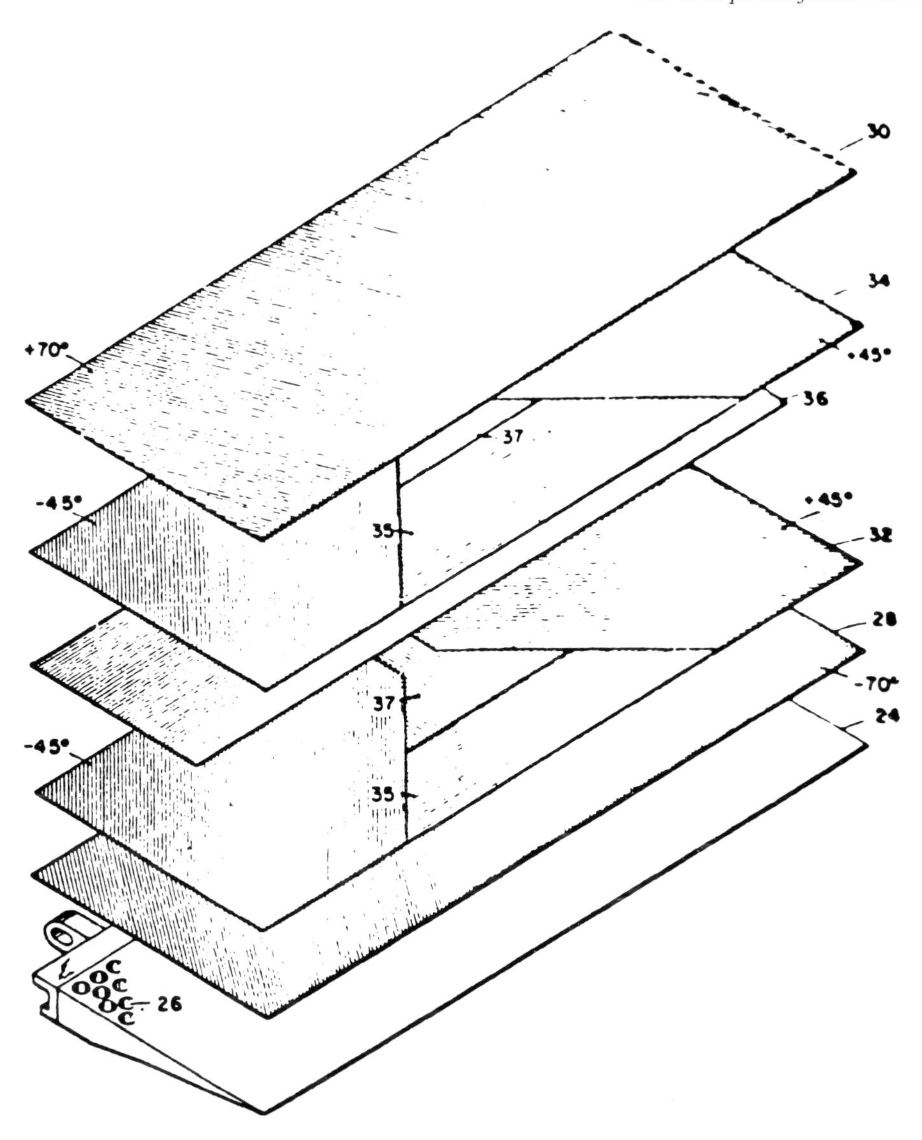

Source: U.S. Patent 3,768,760

Figure 7.3. Stacking of composite prepreg plies in a particular sequence to achieve design goals.

such a procedure through design of a suitable fixture. Components can also be fabricated in individually heated self-contained tools. In this case, heating elements are built into or made part of the mold form. Consolidation is sometimes achieved through the clamping action of sections of the mold.

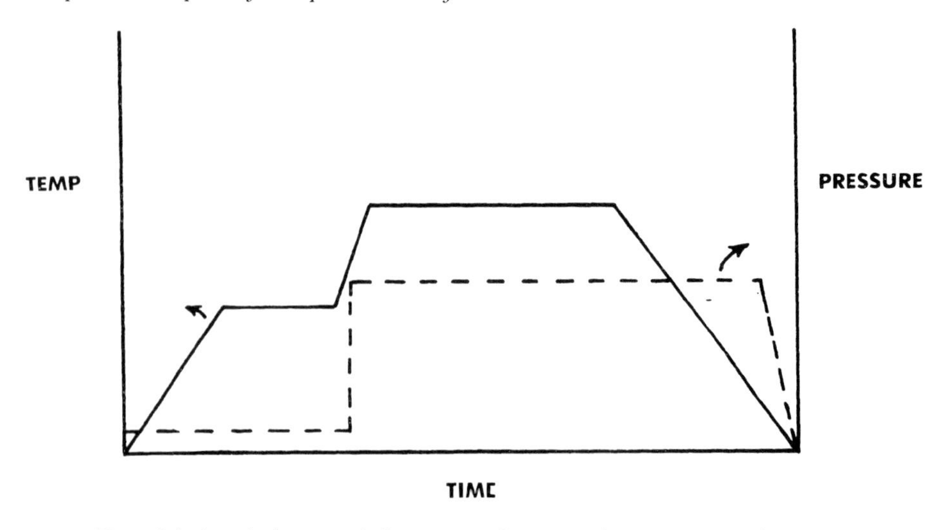

Figure 7.4. A typical cure cycle for a conventional graphite/epoxy composite system.

7.3 Defects

Types of defects

Many types of defects are possible in composite structures. They may be introduced during composite manufacture due to errors in manufacturing procedure. Defects may be initiated during the life of the component as a result of exposure to the service environment, by accumulation of minor damage sustained in the normal use of the composite, from occasional exposure to an abnormal environment, or from a local mechanical overload due to misuse. Examples of types of defects are shown in Table 7.1.

Some defects may grow due to exposure to the operating environment of the

Table 7.1. Examples of defects in composites

Defect	Probable cause
(1) Manufacturing Defects.	
Voids in the matrix	Poor control of the resin
Missed or misaligned fiber layers	Poor layup control
Inclusions and delaminations	Foreign objects, poor thermal controls
Surface damage	Inadequate mold release
(2) Progressive Defects.	
Loss of matrix rigidity	Absorption of moisture etc.
Loss of fiber-matrix bonding	Absorption of water
Hole elongation	Overloads due to bolted construction
Surface degradation	Embrittlement of resin, rain & hail erosion
(3) Defects due to Abnormal Conditions	
Impact damage or delamination	Overload
Crushed core	Impact
Edge damage	Mishandling
Penetration	Battle damage
Surface oxidation	Overheat, lightning-strike

aircraft. This is especially true of delaminations and disbonds which may grow, especially under cyclic, compressive stress conditions. The effect of defects on the strength of metallic components and their service life is covered by the science of fracture mechanics. In metallic components, the conditions under which cracks initiate and grow are understood, and components can be designed to tolerate defects of easily detectable size. Growth of defects during service can often be predicted. Techniques for monitoring critical components can ensure the safe operation of damaged components until such a stage that defects can be detected and well before their size becomes critical. This is not the case with composite components. Fracture mechanics is not as well defined for composite materials. In composites, the significance of a defect of a certain size must be determined by coupon testing under the static and fatigue loads predicted by the component design.

A defect may affect the component in two simple ways. First, the defect may degrade the strength or stiffness of a component to a level below its design limits. A defect of this nature must be detected and the component repaired or replaced. Of more concern, are smaller defects which do not degrade mechanical performance, but which will grow under service conditions to a point where residual strength or stiffness is unacceptable.

The problem of defect significance must be considered in terms of inspection techniques available and the fracture behavior of composites. Several recent symposia have addressed these problems together [11–13].

Non-destructive inspection

Three general types of inspection are regularly carried out. The first type involves inspection of new or reworked components at the manufacturing facility. The second type involves inspection of used components at depots where the aircraft are overhauled, and where complex instrumentation is available. The third type involves inspection on, or near, the aircraft at the operation base. Here the equipment must be portable and operator expertise may not be as high as at depot or manufacturing facilities.

The most commonly employed techniques are based on visual inspection, sonic transmission, ultrasonic transmission, radiography, and eddy-current absorption.

Field inspections on, or near, the aircraft frequently rely on visual inspection for impact damage, fastener hole elongation, local blistering, or other macroscopic damage. This is augmented by simple coin-tapping, which gives a good indication of the quality of the surrounding area.

Acoustic techniques.
Coin-tap will frequently detect delaminations, disbonds, matrix softening, and some cracking. It is, however, particularly subjective, and depends critically on operator experience. Some automation of this technique has resulted in development of instruments to allow more objective interpretation of the transmitted sound quality.

Some attempts have been made to use acoustic emission techniques in NDI procedures [14]. This method uses a number of detectors designed to detect the minute sounds generated by failure of individual fibers, or slight movements of a growing crack, during proof-stressing of the component, or during normal use. This technique is still in the developmental stages for composites, although some success has been reported with metallic and composite parts.

Ultrasonic techniques.
Ultrasonic techniques can be used in the field [15]. The majority of available instruments use A- or C-scan techniques where an ultrasonic pulse is transmitted into a sample by a piezoelectric transmitter. Major discontinuities such as surfaces, interfaces and defects reflect the transmitted pulse and the reflected echoes are received by detectors tuned to the transmission frequency. Often combined transmitter/detector probes are used. The signals are processed electronically, to give an output related to the distance between transmitter/receiver and the discontinuity.

Ultrasonic scanning of the components is the NDI technique which is most widely used for the determination of defects in field and factory environments. Simple hand-held units are readily available for field use. Using these units usually involves moving a transmitter/receiver probe, or a pair of probes, over the surface of a composite component. The probe, or probe-pair, may be set-up to look for reflected pulses received before the reflection from the back-face of the component. Any flaw or defect of adequate size will give rise to a signal which may be indicated on a meter (amplitude related to the defect size), or on an oscilloscope screen (reflection size and position relative to the backface, may be seen). Detailed maps of defect size and location may be constructed on the component, manually.

In factory environments, the need for mechanical contact between probe(s) and component may be eliminated by immersion in a bath or by use of a continuous stream of water. Automatic scanning and complex signal processing will allow the construction of maps of defect locations in three dimensions and will allow some classification of defect type.

Other ultrasonic methods determine the velocity of sound in the component and the dissipation factor. Values for these constants can provide information on the modulus of the material and its state of degradation.

Radiography.
X-ray radiographs are often used to examine composite components [16]. Because the fibers and organic matrices have similar absorptivities, the contrast in the radiographs is lower than in metal components. The use of lower energetic X-rays can improve contrast. Defects which are directly connected to the surface of the component can be detected more readily if enhancement techniques are employed. Here a liquid which absorbs X-rays readily is allowed to penetrate into the sample. Typical liquids include aqueous solutions of heavy metal salts and some halogenated organic derivatives. This method is very useful for determining the extent of impact or penetration damage.

X-radiography is also widely used to examine composite to metal joints, composite porosity and to search for liquid water in honeycomb structures.

Neutron radiography shows some promise, particularly for detection of water, or areas of unusually high resin content. Problems associated with transportability of the neutron source, restrict its use at present.

Eddy-current techniques.
Methods using eddy-currents induce an electric current in a conducting material by holding a radio-frequency transmitting probe in close contact with the surface [17]. Study of the properties of this current has proven useful in the detection of flaws in metal components and many instruments are available for routine use. Only graphite composites can be studied using this method. Even in these materials the signal generated is weak and variable.

One of the most important stages of repair is to establish the area that is defective. Of the techniques described above, ultrasonic inspection using C-scan methods and X-radiography, preferably using an enhancement technique, are the best methods to ensure that an adequate repair can be considered. A careful study must be made, to decide whether the material is generally sound and can be repaired by a superficial method such as resin injection to rebond delaminations and debonds, or whether fiber degradation has occurred, in which case new structural material must replace the lost strength.

Growth of defects

Many studies have been made of the growth of defects in composites. Generally, few cases of growth of any type of defect have been observed, except under conditions of compressive load, when individual fibers debond and fail in a buckling mode. Hot, moist environmental conditions accelerate this failure, probably by increasing the rate of debonding and by reducing the matrix modulus and support for the fiber. Small areas of delamination caused by low energy impact can propagate under cyclic compressive loads to give rise to larger areas of physically degraded material.

7.4 Repair materials

The repair design must include a consideration of the materials needed to perform the task. The objective of a repair action is to restore the physical and mechanical properties of a composite component and allow it to function in its operational environment. It is desirable to use repair materials which are similar to those used in the original structure. For example, repairs performed on damaged graphite-epoxy (gr/ep) composites typically utilize epoxy resins to fill delaminations and gr/ep patch materials to repair large damage. Other patch materials may also be used. The use of titantium foil patches to repair gr/ep has been successfully demonstrated [18]. Similarly, the adhesives selected must possess properties comparable to those used in the original design.

Selection of the adhesive is a particular problem in repair of composites. For repairs conducted on, or near, the aircraft, full support of the component in an

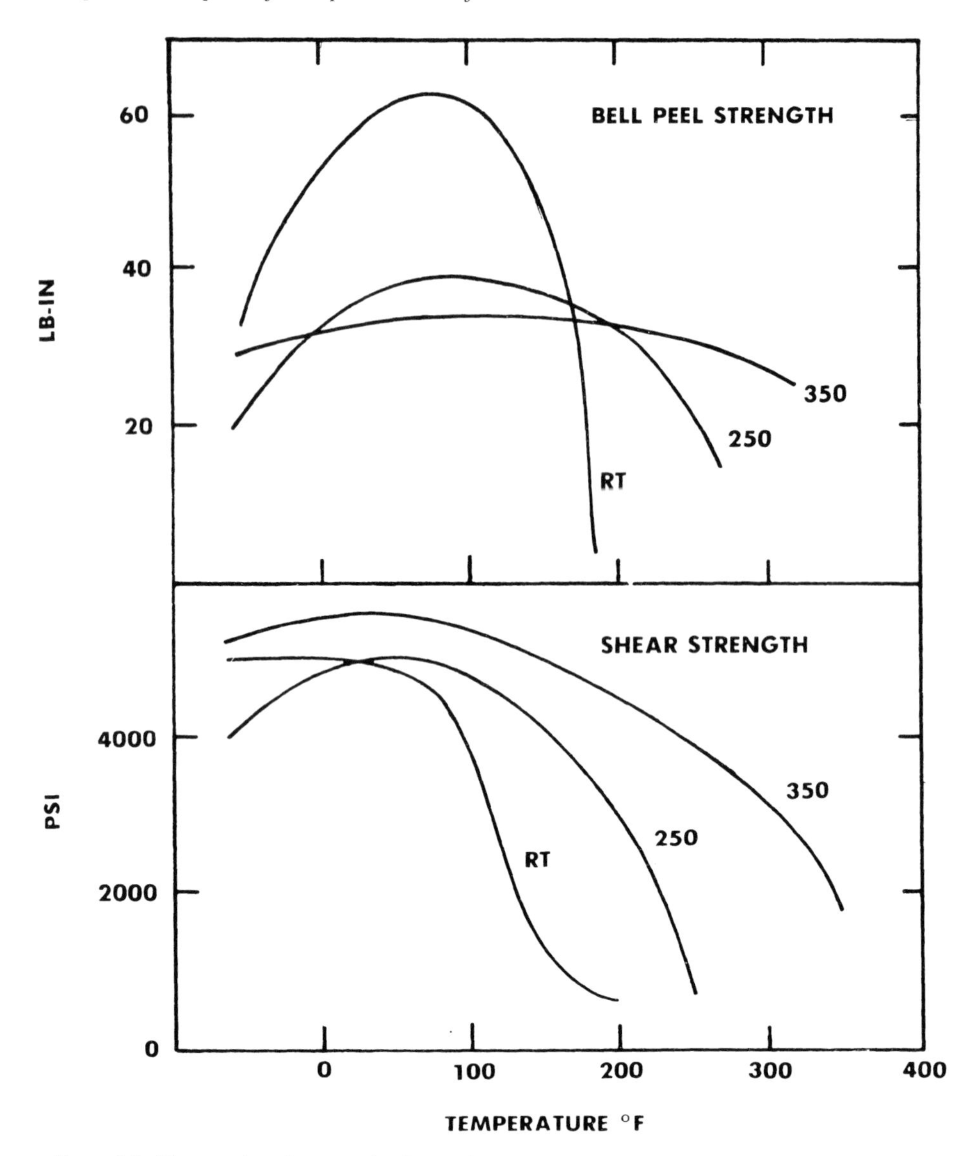

Figure 7.5. Shear and peel properties for typical ambient, moderate, and high temperature cured adhesives.

appropriate fixture is not possible, and the cure temperature of the adhesive must not be over the maximum exposure temperature of the component. This can seriously limit the choice of adhesives for particular applications, especially in repair of high temperature composites such as graphite/bismaleimide or graphite/polyimide composites. In general, adhesives which are resistant to exposure to high temperatures tend to have low resistance to peel forces at low temperature. These peel stresses must be considered along with the adhesive shear stresses in patch design. Figure 7.5 shows shear properties and peel properties for typical ambient, moderate and high temperature adhesives.

There are a number of alternative materials forms available for gr/ep repair applications. The forms include precured patch, co-cured patch, and wet laminate. These types of repairs are shown, along with some advantages and disadvantages of each approach, in Table 7.2 [19]. In this case, AS/3501-6 is the gr/ep used in the component which is being repaired. These forms are used to produce step, scarf, and external patch repairs. A number of programs have examined the use of external patch and scarf repair concepts which utilized precured, cocured and wet layup patch material forms [20].

The heat cure cycle required in the bonded repair process is defined by the material selected and the specific application. The repair must function in the thermal environment of the component. If the part is exposed to mild ambient temperatures, then low temperature curing polymer systems may be used. If, however, the part is exposed to high temperatures, some type of heat cured system is required to assure thermal insensitivity in the patch.

A good deal of effort has been directed towards the development of injection

Table 7.2. Material forms for patches[19]

Material form	Method of application	Advantages	Disadvantages
1. Precured graphite/epoxy patch (e.g. AS/3501-6 with 0/+45/90 ply orientation protected by peel ply.	Remove peel ply. Cover laminate patch with film or paste adhesive. Clamp or hold in place with vacuum until cured.	Excellent for flat surfaces. Incorporates high strength laminate into the repair structure.	Not applicable to contoured surfaces.
2. Precured graphite/epoxy patch (e.g. AS/3501-6) with 0/+45/90 ply orientation covered with ambient temperature or low temperature curing film or paste adhesive that is protected by polyethylene film.	Remove protective film from adhesive. Activate the surface of the adhesive with the appropriate activator. Clamp or hold in place with vacuum until cured.	Same as above. In addition, the adhesive will not have to be applied separately.	Not applicable to contoured surfaces. Anaerobic curing adhesives will require special packaging to prevent cure in storage.
3. Dry woven graphite fabric coated with ambient temperature of low temperature curing liquid resin.	Prepare a wet layup over the damaged surface area. Use film or paste adhesive or resin bonds.	Simple operation. Uses well-established boat repair technology. Useful for both flat and contoured surfaces.	Cured laminate will almost certainly have a high resin content and a low fiber volume.
4. Low temperature curing graphite prepreg tape of fabric.	Position uncured prepreg over damaged surface using film or paste adhesive. Apply heat as required. Use vacuum bag pressure during cure.	Useful for both flat and contoured surfaces.	The mechanical strength using vacuum bag cured laminates will be less than that from the autoclave cure of the AS/3501-6 precured patches.

techniques for filling small delaminations. Low viscosity epoxy adhesive are usually used in this application. Again the requirement of cure cycle depends on the specific thermal environment of the particular component. Vacuum assisted injection techniques have shown an improvement in wetting of the delamination.

Other composite materials used in the aerospace industry include glass reinforced epoxy, Kevlar reinforced epoxy, and graphite polyimide (gr/pi) systems. The Kevlar and glass composites are used most widely in rotary wing aircraft. The repair concept for these materials is again to use similar materials as those used in the original construction.

The most significant recent advance in composite technology has been the development of high temperature composites based on thermally stable polymers such as polyimides. Because of the application, repair materials must also be thermally stable. The conventional approach is to use polyimides as repair materials for damaged components. The nature of these materials necessitated high pressure, high temperature repair processes. Repair techniques for gr/pi have been successfully demonstrated [21]. Developmental work aimed at reduction of the processing pressure requirement is being performed. Examples of the directions taken include modification of the basic polyimide chemistry and formulation of high flow thermally stable polymer systems.

The repair environment will also restrict the choice of materials for use in repair operations. At a rework facility, major repairs are possible, but at a forward or field position the limits on repairability are much more restricted. The rework facility is equipped with freezer storage for materials, autoclave capability for processing, and tools designed for repair applications. Also the level of expertise available at a rework facility is higher than that available in the field. Thus, parts which are easily detachable from the aircraft can be repaired under ideal autoclave conditions. Also, manufacturing materials, both pre-pregs and adhesives, can be used in the repair procedure. The rework facility is capable of performing all repairs required for the aircraft.

The field repair environment is more severe. No freezer storage may be available, and the materials used should be capable of ambient temperature storage. There is no access to autoclave facilities and thus repairs must be performed with heating blankets. Compaction and conformational pressure is usually limited to vacuum bag pressure of 100 KPa. Also, the personnel and tooling available restrict the repairs to external bonded patches. In critical structural regions, pre-cured or pre-staged patches must be used. In less critical areas, wet-impregnated repair patches may be used.

Additional materials which may be required in repair applications include potting compounds and sealants. The first step in the application of a bonded repair after substructural damage has been repaired is the preparation of the bonding surface. For repairs on monolithic skins, preparation involves clearing the damaged area of debris and then filling the cleaned hole with a potting compound. The potting compound most frequently used is a filled epoxy resin.

A sealant may be used to prevent moisture intrusion around the periphery of the patch. This material is also used around "wet" installed fasteners to prevent corrosion and moisture intrusion. The use of a suitable sealant is important when repairing "wet" areas of the aircraft where fuel leakage must be prevented.

There are a number of efforts currently directed towards repair material development. The aim of these programs is to simplify the repair procedure and to reduce the total time required for the process [19, 22]. Examples of ongoing programs include work in ultraviolet light curing resins for glass composite repair, and work on the development of thermoplastics and techniques for their use in repair applications [23, 24].

7.5 Composite repair concepts

Composite repair has extensive application in modern aircraft in which increasing consideration is being given to composite constructions for various components. When discussing repair, it is necessary to define realistic repair criteria and then develop concepts which will meet the established requirements. It is important to simplify the repair procedure as much as possible in terms of equipment, time and skill required.

Criteria for composite repair can be thought of in two categories. First, the repair must function in the mission environment of the aircraft. That is, the thermal stability and strength are critical for operation. The second set of criteria result from a consideration of the constraints of the repair environment such as equipment availability and space.

Consideration of necessary strength for repairs usually requires design ultimate tensile and compressive strain levels in monolithic skins and the same strain level for both faces of bonded honeycomb construction. The repairs must not result in any reduction of stiffness but small local increases are permissible. Care must be exercised in not adding excess weight in a specific area particularly when repairing honeycomb where flutter may be a concern. Aerodynamic compatibility must be considered even though external surfaces need not be smooth. Generally, steps on outside patch surfaces are not to exceed 2.5 mm. Spectrum fatigue loading is another factor. The load levels and type and number of load applications are predicated on the requirements of the specific aircraft.

Two general approaches can be taken to the problem of composite repair, namely bolted and bonded repair procedures. Both have their place in repair of aircraft composite components. The factors which determine a particular approach are: the specific component, laminate thickness, damage size, accessibility, load requirements and repair capability.

7.5.1 Bolted repairs

Bolted repairs can be used when thc laminate thickness exceeds 8 mm. Thinner laminates cannot withstand the bearing loads induced by the bolts. External bolted patches can be applied from one side or from two sides with a backing plate. These plates can be inserted through the damage to gain access to the back side. If the plates are thick, they can also carry load. The patch must be thick enough in some cases to accept flush head fasteners. There is a practical limitation on patch thickness and size based on aerodynamic considerations. The patch may have beveled edges to improve aerodynamic conformability. The limit on damage size which can be repaired is approximately 75 mm [25].

Another approach to bolted on repair is a flush type patch. In this case, the damage is "cleaned up" and a section is inserted which is now flush with the surrounding undamaged area. A doubler is mandatory with this approach. Fasteners can be applied through the patch to the doubler as well as from the undamaged area into the doubler. One of the difficulties associated with the flush type metallic patch is potential difficulty with limited back side access to install large doublers. The larger doublers arise from the great number of fasteners required with such a repair.

Bolted repair techniques use aluminum and titanium as patch materials. Aluminum is lighter than titanium and offers less of a weight penalty. It is also, however, significantly electronegative or anodic to Graphite. This leads to possible corrosion problems. Aluminum patches must be physically separated from the underlying composite by scrim cloth or sealant or both. Titanium will not corrode in the presence of graphite. Titanium presents additional difficulty with regard to machinability and formability. Specific titanium alloys are being made available which are easier to form or shape.

The design of the patch used in any repair application is based on finite element analysis. The number, size, and installation pattern of the bolted fasteners is determined by the analysis. NASTRAN is one such finite element analysis method. In one case the NASTRAN finite element models used flat anisotropic rectangular and triangular elements for the panels, isotropic rectangular and triangular elements for the metallic repair patches and backup plates and one-dimensional beam elements for the fasteners [25]. In this case, bending of the panel and patches was ignored and the repair configuration was idealized as planar sheets interconnected by flexible terms. In this fashion, complex three-dimensional behavior of the joint was reduced to two dimensions.

A number of different types of failure modes have been analyzed for bolted repairs. Some of these are: net tension at the repaired hole, laminate bearing and tension interaction and fastener shear failure. Design allowables have been developed for each of these failure modes. Specific programs exist [26] which can predict edge of hole strain and bolt load distribution. Using a combination of design allowables and such program predictions, bolted repairs can be evaluated to verify their ability to withstand particular load levels prior to failure. Bolted patch repairs are expected to fail when the predicted edge-of-hole strains exceed the strains developed by an unrepaired test specimen at failure.

The allowable shear load that can be transferred by the fastener in a titanium patch repair depends on the countersink depth in the titanium. A large percentage of the bearing load is carried by the cylindrical portion of the hole through the patch and the remainder is supported by the countersink portion. When the countersink approaches full depth, the fastener head may be guillotined before it can develop the single-shear allowable strength. The fastener loading in bolted repair can be different depending on the circumstances. Blind fasteners are loaded in single shear. The use of a thin (relative to patch) backup plate with average holes also results in single shear. Using thicker backing plates (relative to the patch) and close tolerance holes changes the situation to one in which the fasteners are loaded in double shear.

Bolted repairs are not without problems. The drilling operation can be time-consuming and deleterious to the laminate integrity. The drilled holes could be considered as the introduction of additional damage into the area surrounding the original damage. Delaminations in particular could arise from improperly drilled holes. Bolted repairs cannot be used over honeycomb areas. The additional holes would further degrade the region and honeycomb doesn't provide a suitable attachment medium for fasteners. Non-flush fasteners, depending on the number and pattern, could also provide aerodynamic resistance on external surfaces.

7.5.2 Bonded repair

When considering bonded composite repair, it is important to categorize the severity of damage and the implications on the operational capability of the aircraft. For the purpose of simplicity, the types of repairs will be divided into three categories; nonstructural, secondary structural, and primary structural repairs.

Nonstructural repairs.
Nonstructural damage includes gouges, scratches, dents, or other defects which are confined to the surface of the composite laminate. Such damage types can be dealt with by utilizing "cosmetic" repair techniques. In such instances, filled two-part adhesives are frequently used to repair such damages. Generally, the two-part system is one which can be stored under ambient conditions. It should have a suitable pot life to allow enough time to apply it and fill the particular damage. Abrading the damage surface and solvent wiping aids in enhancing adhesion. Although this is not a critical application, such surface preparation helps ensure some degree of repair longevity. A spatula or similar device aids in spreading the resin to match the surrounding surface. Sufficient time should be allowed for the resin to harden, although complete cure is not particularly necessary before flight. A variation of this approach is sometimes used in which chopped fibers such as glass or graphite are added to the resin. This is done when the damage is deeper or encompasses a larger area. Such fibers impart a higher degree of mechanical strength to the repair. The most important consideration for these repairs is application of the resin to match the aerodynamic contours of the area.

Another type of nonstructural repair which could be considered is the case of holes in laminates which are of little structural significance. These are not serious defects either because of the small size of the holes or the location in the aircraft. An example of an innocuous location could be a hole in an access door. The procedure in such a case would be first to clean up the area by suitable abrasion and wiping. The next step would be to fill the hole with a two-part or one-part paste. A backing plate may be used to aid in filling the hole and preventing excess resin from flowing into the inside of the component. The resin could then be applied without concern for material falling through the hole.

At times, small damage to composite fuel tanks is encountered. In this case, the primary concern would be stopping the leakage of fuel. The repair applied must be able to withstand the positive pressure which exists in the tank. If the damage were extremely small, the puncture could possibly be filled with polysulfide sealant. In most cases, this would have to be preceded by first draining the tank and cleaning the damaged area. In the case of larger damage, the holes would be filled with a sealant or epoxy system. After application of such a material, a section of cardboard or suitable material could be pressed against the exterior of the tank until the resin or sealant was sufficiently rigid. Such a substance could also first be applied to a small section of rigid laminate and cured in place with the backup material. In instances where more structural reinforcement is required a wet patch could be applied after the hole has been repaired. A suitable amount of cloth, usually glass, could be impregnated with a low viscosity resin and stacked over the specific area. Additional consolidation could be achieved by applying weight to the patch, by clamping, or with a vacuum bag. If required, heat for cure could be supplied by a heating blanket or heating lamp.

Secondary structural repairs.

Secondary structure can be thought of as those components on which flight dependence is not critical. Access doors, speed brakes or covers are examples of secondary structural components. The loss of these components in flight would not cause loss of the aircraft. These structures usually consist of thin laminate construction less than 2.5 mm thick. Thus, strength is not a critical factor with such components. The components will usually possess more 45° or other off-axis plies as compared to 0° plies. The exact ply orientation will be dependent on stiffness considerations particularly in the bending or torsion mode. The primary objective for repairs in these cases is restoration of stiffness or stability. These repairs are usually thought of as on-aircraft repairs because they are of a less critical nature. It is usually necessary to fill the actual hole in order to provide a good bonding surface for the patch. Typically, such repairs are 16 plies or less, and can be expected to recover 25%–60% of unflawed laminate strength. Precured, cocured, and wet patches, can be used for such repairs. Precured patches with 2-part or film adhesives have been used with success in such applications [27]. Cocured patches can be used with a suitable film adhesive.

In the past, radomes have frequently been repaired using a wet impregnation technique. Heat and pressure required for cure are applied with a heating blanket or heating lamp and vacuum bag. Such patches have to be relatively thin since the eccentric load path results in severe bending in the patch and peeling stresses in the adhesive and composite. Compressive axial loading can bring about out-of-plane bending and reduce buckling stability. Quasi-isotropic layups (equal properties in the plane) are usually used to reduce application and layup errors.

Resin injection repairs are a different type of repair used mainly for secondary components in which minor disbonds or delaminations exist. Manufacturing flaws, those due to a lack of adequate composition or contamination during fabrication, will result in a smooth, difficult to bond surface. Since adequate surface preparation cannot be assured for internal surfaces, it is not feasible to attempt this type of approach with such a defect. However, delaminations caused by impact usually present a roughened surface which is easier to bond. The problem with injection into delaminations is that they can be very acute and difficult to fill. Also if the defect is not penetrated, the resin cannot enter the void; if the hole is too deep, damage will occur below the void, and the injection may be inefficient. An impact can produce multiple delaminations. Finally, prolonged heating of the component prior to the injection may be required to remove moisture. In an injection procedure, a number of injection and bleeder holes are drilled to the depth of the void. The resin is pre-heated to reduce its viscosity. The resin is injected until excess resin flows from the adjacent bleeder holes. This procedure is repeated until all such holes have been treated. The holes are then temporarily sealed with a layer of protective tape. Finally, pressure is applied to the repair area to improve mating of adjacent regions and to improve or maintain contour. The resin usually gels at room temperature and can be post-cured in the region of 150°C. A similar procedure can be attempted for disbonded joints. Injection usually does not appreciably help joint strength and would be used mainly to prevent moisture ingress. In such cases, injection is most warranted where the disbond extends to a free edge where ready access is available. A sealant or low modulus adhesive can be used to seal the area.

Primary structural repairs.
Primary structure is that part of the airframe which is critical for flight. Failure of a primary component results in loss of the aircraft. Bonded repair of such components is more involved than the previous methods since the repair must be capable of effectively transferring more load. There are usually two types of bonded primary structural repairs; a flush scarf repair and an external patch repair.

Bonded flush type repairs are used where aerodynamic smoothness is required. Such repairs provide the maximum joint efficiency and are applicable where load concentrations and eccentricities must be avoided. Three examples of this type of repair are a full thickness flush repair, a partial thickness flush repair and a flush repair with internal doublers. Such repairs have been found to achieve 60% of unflawed laminate strength. It is assumed that the joints would be scarfed and plies matched between the repair and surrounding area. These repairs require careful preparation and are time consuming regarding machining and application. Scarf repairs do, however, give higher load carrying capability than other methods of repair.

Scarf bonded repair concepts have been developed for repairing monolithic laminates up to 10 mm thick. There are a number of criteria for such a repair. First the damage size is limited to an area approximately 150 mm in diameter which can encompass one or two substructure spars or ribs. These repairs are assumed to be made in place and no access is available to the back side of the damage. The thickness of the skin can vary, particularly the inner moldline which is contoured for ply dropoffs and buildups. The repair also has to develop the full ultimate

JOINT CONFIGURATION	RELATIVE JOINT STRENGTH
 A. BASELINE	100%
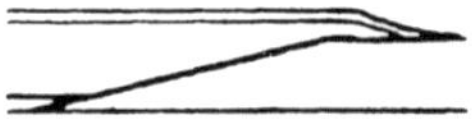 B. PRECURED SUPPORT	94%
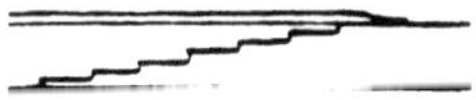 C. STEP LAP	87%
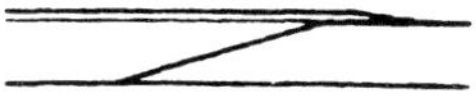 D. PRECURED PATCH	82%
 E. NO INSIDE DOUBLER	78%
 F. STEEPER SLOPE	71%
 G. CONTINUOUS PLIES	58%

Figure 7.6. Relative hot/wet strengths of various scarf repair configurations.

design allowable strength of the skin. Projections on the air side of the structure are to be minimized and the repair must be amenable to easy application. Figure 7.6 shows a number of variations in bonded scarf joints for thick monolithic skin repair. The relative strengths of wet specimens at 150°C are shown as percentages of the baseline configuration. The second configuration (b) used a precured support and the relative strength dropped to 94% of the baseline (a). A step lap joint was used in (c) and the strength decreased to 87%. Configuration (d) used a precured patch with an additional slight decrease in strength to 82%. An inside doubler was omitted (e) which resulted in a relative strength of 78% of the baseline.

Configurations (f) and (g), using a steeper scarf angle and continuous plies on the top surface resulted in relative strengths of 71% and 58% respectively of baseline. In all cases, the plies in a scarf joint must be carefully cut to match the surrounding laminate orientation. Usually in a thick laminate, there are a substantial number of 0° plies necessary in order to achieve desired mechanical properties. This requires that the repair must maintain the same proportion of 0° plies. A typical orientation might contain 50% 0°, 40% 45°, 10% 90° plies in a number of different stacking sequences. The fit within the prepared area must be very good to insure full attainment of desired properties. The usual preparation in this case involves close tolerance matching to achieve the desired scarf angle and tolerances. Solvent wiping after machining of the scarf is required to remove loose particles and/or debris.

There are a number of advantages and disadvantages with the types of scarf repairs listed above. A precured scarfed patch is easy to install against the prepared surface, it provides no interference with the substructure, and it gives excellent aerodynamic smoothness. The ability to achieve the desired scarf angle tolerance is the chief drawback to this approach. A precured stepped scarf patch has less critical scarf angle tolerances than the constant slope patch. In the stepped scarf patch the variation in step depth has a critical effect on bondline thickness and repair quality. Peaks occur in the shear stress at the end of each step making steps less effective than a scarf joint. Also, variations in cured thickness of the laminate can affect the depth of steps, so that a step may not have the intended ply orientation exposed. Another scarf type previously discussed embodied an inner doubler on which the precured patch fitted. The disadvantage of such an approach is that the fit of the inner doubler may be difficult. It is difficult to seal the region where the substructure meets the edge of the cutout. A cocured scarf requires a tapered doubler to close the hole and provide support. A shear stress concentration results at the tip of the support plate if no internal doubler is provided. All of these approaches can use a vacuum bag or autoclave procedures to affect the repair.

The other approach to bonded repair of monolithic skin is the use of external patches. In this case, the load is carried over the damaged area. One variant of such an approach would be to cocure the prepreg and adhesive in one operation. The reinforcement for the prepreg can take the form of tape or cloth. The separate prepreg sections are cut to various sizes and stacked over the filled in damage region. The various patch configurations are shown in Figure 7.7. The smallest section must cover the damage and increase progressively either up or down.

This sort of stacking sequence is necessary to "fair in" the load and avoid large stresses at the end of the patch. Adhesive, usually in film form, is cut to the size of the largest section of prepreg material and placed between the patch and damaged region. Prior to placement of prepreg or adhesive, it would be necessary to clean up and fill in the damage with a quick setting thermoset. Cocured patches have an added advantage in that they can be used on curved components and will conform to the curvature during cure.

Precured patches are also used in bonded repair procedures. The patches can be of varying thickness dependent on the requirements of the damaged laminate. Precured patches offer an advantage over prepreg use in that storage is not a

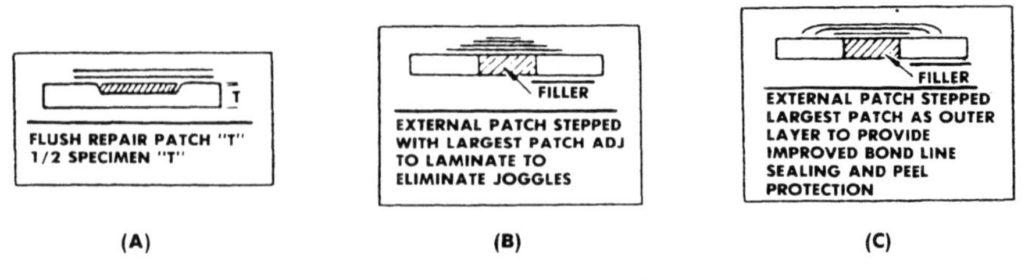

Figure 7.7. Patch configurations used for composite repair.

constraint. Adhesive is used between the patch and base material and must be suitably cured. Storage of the adhesive must be considered. As with cocured bonded repairs, the damage is filled in up to the surface of the surrounding area to provide a suitable base for bonding. Precured patches must be applied over surfaces which are either flat or only very slightly curved to avoid lifting off as a result of high peel and shear stresses at the patch edges. All the previously mentioned bonding procedures require some measure of elevated temperature cure with some means of compaction to achieve strength at appropriate operating temperatures and suitable integrity in the resin materials. These type of repairs are used where a significant portion of load carrying capability is desired to be restored to the particular component, or where repair durability is an important consideration.

Both external patch repairs and scarf type repairs can also be utilized in thin skin structures such as stabilizers. Such components might consist of composite skins 2.5 mm thick adhesively bonded to substructure or honeycomb core. The maximum repairable damage which can be considered in such cases is approximately 150 mm in diameter. Sealing is an important consideration of honeycomb repair to prevent corrosion and moisture ingress. Before the damaged skin can be repaired, it is necessary to repair or replace the damaged core. Two approaches are generally used: either replace or stabilize the damaged core with potting compound, or replace the damaged core with a bonded in section of a similar type core. It is usually preferred to use a honeycomb plug rather than potting compound since the heavier potting compound may cause failure because of increased local stiffness. The potting compound can be cured in place and then sanded prior to bonding on the skin section. In the core replacement approach it is first necessary to remove the damaged core. This is accomplished with a router, a knife blade, or a hole saw. The replacement core plug should be of the same density and cell size, and if metallic core, the same alloy. A higher density smaller cell size honeycomb can also be used. Figure 7.8 shows a typical composite-faced honeycomb repair. Frequently, it is necessary to stabilize the core for machining. If proper machining techniques are used the core can be machined without being filled. It is difficult, however, to hold the core in place during the machining operation. Final shaping can be performed with light sanding; although sanding might distort cell walls which would have to be straightened prior to bonding. After machining and/or sanding, the surrounding core should be properly degreased with a suitable solvent. A foaming adhesive can be used to bond the new section of core in place.

A number of approaches can be used to repair the skin once the core has been

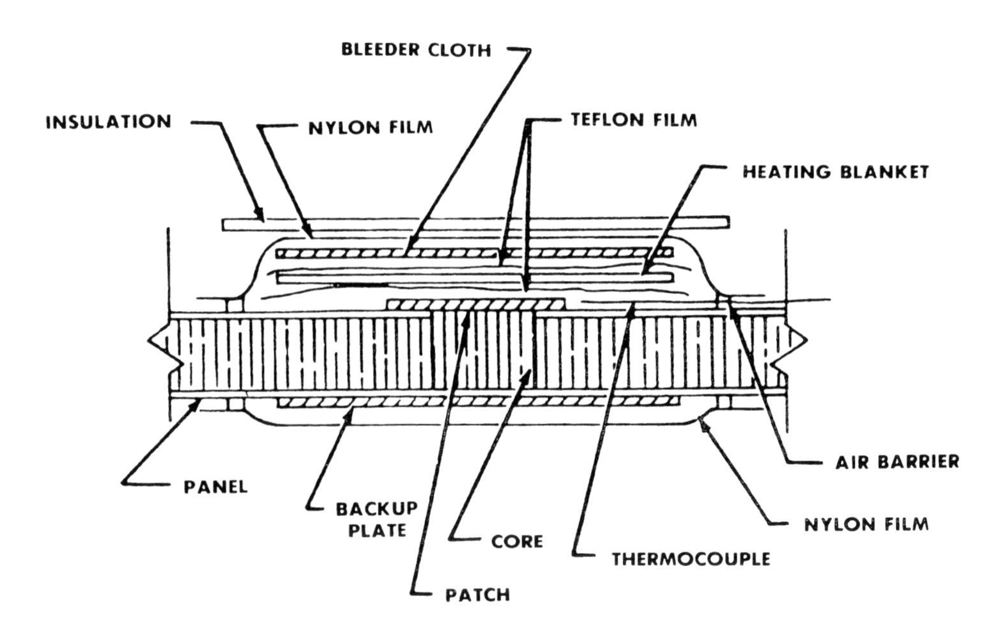

Figure 7.8. A schematic of a typical composite faced honeycomb repair.

repaired or replaced. One type involves using a layer of adhesive, followed by a layer of fiberglass to reduce stress concentrations, and a layer of adhesive followed by alternating layers of precured composite and adhesive. The size of the layers get progressively smaller with increasing height above the skin surface. This particular approach is used with film adhesive and eliminates mechanical attachment and provides good formability to curved surfaces. A disadvantage of this approach is that the patch needed to repair thick skins is very thick which results in eccentricity of the load patch, more weight and poor aerodynamic smoothness. An alternate approach would use a precured or cocured patch with one layer of film adhesive at the skin interface. A cocured patch would provide much better conformance to the original surface than a precured patch. It would be advantageous in a precured patch approach to use a splash of the moldline as a tool to obtain better conformity.

Another way to repair the skin would be to use a scarf joint with a precured or cocured patch. External plies would be part of the patch and would extend like a doubler beyond the scarf area. The adhesive would correspondingly accommodate the scarf joint and overlap the surrounding skin to bond the external plies to the skin. The obvious disadvantage of such an approach is the machining required to create the scarf angle. A variation would be to step the angle and proceed as above with a precured or cocured patch and one layer of adhesive. An advantage could be machining the steps might be easier if router templates were available; a disadvantage would be crucial fit-up required with a cocured patch. All the sandwich type repairs enumerated would be amenable to a heating blanket vacuum bag approach for curing.

7.6 The effect of moisture on bonded repair of composites

Organic matrix composite materials can absorb an appreciable weight of water. The equilibrium amount absorbed depends, primarily, on the partial pressure of water in the surrounding environment. It has been shown [28, 29] that the equilibrium weight fraction of water absorbed in a composite (M_c) is given approximately, by the relationship

$$M_c = \frac{a\phi}{w_r} \tag{1}$$

where M_c = equilibrium weight fraction of moisture in the composite.
a = constant depending on the matrix
ϕ = relative humidity in the environment
w_r = weight fraction of resin in the composite.

Values for the constant, a, are generally 0.01 0.03. Moisture concentrations are then about 3–5% for pure resin and 1–2% for composite at a relative humidity of about 80%.

When the damaged, moist composite is subjected to a repair technique using heat-cured adhesives, the environmental partial pressure due to moisture will be very low. Moisture will then diffuse out of the composite at a rate determined by the appropriate diffusion constant. This moisture will add to moisture already present in the liquid or film adhesive.

If the hydrostatic pressure in the adhesive film is lower than the partial vapour pressure of the moisture, it will then volatalise and cause extensive void formation in the bond line.

To predict the conditions under which such voiding will occur, it is necessary to know the diffusion constant of moisture in the composite. To a first approximation, it has been shown that for the one-dimensional case the rate of diffusion is given by the relationship:

$$\frac{\partial C_x}{\partial T} = D_x \frac{\partial^2 C_x}{\partial X^2} \tag{2}$$

where C_x is the moisture concentration in the composite at a depth x from the free surface and D_x is the diffusion constant in the x-direction. This is one definition of so-called Fickean diffusion. The extent to which moisture diffusion in composites can be described as Fickean has been discussed [30–32]. During absorption of water a two-stage process has been described [32] which gives rise to deviations from Fickean behavior. This may be due to the formation of microcracks in the matrix. The diffusion constant can show some dependence on moisture concentration [31]. Desorption, however, appears to be well described as a Fickean process [30] and desorption is most applicable during the repair procedure.

In order to prevent void formation in the adhesive it is necessary either to prevent access of moisture to the bondline by careful drying of the composite and adhesive prior to use or to control the pressure in the bondline during the repair procedure to prevent volatilisation of moisture present.

Methods of calculating moisture content in a composite exposed to drying conditions have been developed [28]. The following approximation can be derived from Fick's Laws with the appropriate boundary conditions:

$$M = \left\{1 - \exp\left[-7.3\left(\frac{D_x t}{s^2}\right)^{0.75}\right]\right\}(M_m - M_i) \tag{3}$$

where M = moisture content after drying (in wt%)
M_m = equilibrium moisture content (see equation)
M_i = initial moisture content
s = maximum distance for diffusion (in mm)
= thickness if moisture loss from one side
= half thickness if moisture loss to both sides
t = time (in sec).

The diffusion constant for epoxy composites has been found [30] to be given by the expression

$$D_x = (1 - \sqrt{v_f/\pi})\, D_0 \exp^{-c/T} \tag{4}$$

where

D_0 and c = constants
T = absolute temperature
v_f = fiber volume fraction.

The constant D_0 is about 4 mm^2 sec^{-1} and the constant c is about 5200 °K for matrix resins normally encountered in structural applications.

These equations may be used to show that a 12.5 mm laminate of initial moisture content 1% exposed to a temperature of 60°C for ten days at a relative humidity of 10%, will have a final moisture content of 0.97%. Even at 99°C and 1% relative humidity, the moisture content will be over 0.80% after 24 hours drying.

Full drying of the composite is obviously not a practical possibility. In practical applications, it is also unlikely that the adhesive can be kept completely dry.

Attempts have been made to calculate the conditions under which moisture diffusion through the laminate can be kept to a sufficiently low rate to minimise the voiding problem [33, 34]. Augl [33] uses a finite difference technique to calculate (using a computer) the partial pressure due to moisture diffusion form a partially dried composite. He suggests that by use of controlled pressure in the bondline, it would be possible to eliminate void formation. A method of calculating bondline pressures during bonding with liquid adhesives has been derived [35] but practical results from the use of these techniques have not been reported. Myhre *et al.* [34] have also suggested techniques to reduce or eliminate voiding by selection of curing conditions associated with partial pre-drying of the composite. These authors report on the use of an instrument for determination of moisture content of the composite prior to bonding.

To date no satisfactory technique has been proposed to allow rapid estimation of the probability of void formation due to moisture during repair of composites.

A possible rule of thumb for calculation of adequate drying times may be available by use of the concept of "drying-depth". If a composite of uniform moisture content is exposed to a drying environment for a short period a moisture profile will be set up through the thickness of the composite. At the skin of the material the moisture content will be equal to the moisture content of the surrounding atmosphere, as the process whereby water is lost is much faster than the diffusion process. In the center of the material the moisture content will be unaffected at short exposure times. An approximate relationship giving the concentration of moisture through the composite may be derived from Fick's Laws [36]:

$$C(z, t) = C_1 + \frac{C_0 - C_1}{2}\left(1 - \operatorname{erf}\frac{z}{2\sqrt{D_x t}}\right) \tag{5}$$

where $C(z, t)$ = concentration of water at depth z from the surface at time t
C_1 = initial water concentration
C_0 = atmospheric water concentration
erf = error function, available from mathematical tables.

This relationship is a good approximation for diffusion at short times, and describes a sigmoidal shape curve abruptly transitioning from the central plateau at concentration C_1 to the equilibrium surface concentration C_0. A good approximation of the depth affected by the drying process is the distance from the surface to the plane at which the moisture concentration is $C_0 + C_1/2$. This distance defined as the drying depth z_d is given by

$$z_d = \sqrt{D_x t}. \tag{6}$$

To use this concept the drying-depth which would be caused by application of the cure procedure to the moist laminate is calculated. For example consider a moist graphite T-300 in resin Fiberite 1034 carbon fiber composite to be repaired using an adhesive curing in 1 hour (3600 sec) at 120°C. The diffusion constants for this system at 105°, 120° and 150° are respectively 3×10^{-7}, 2×10^{-6} and $5 \times 10^{-6}\,\text{mm}^2/\text{sec}$. The drying depth will be:

$$\begin{aligned} z_d &= \sqrt{2 \times 10^{-6} \times 3600} \\ &= 0.085\,\text{mm}. \end{aligned}$$

The component should be dried by a preliminary exposure to drying conditions to give a drying depth greater than this depth by a safe factor. For this example, take a drying condition of 105°C and use a safety factor of two. The time necessary to dry the composite to this level given by

$$\begin{aligned} t &= \frac{4z_d^2}{D_x} \\ &= \frac{4 \times 0.085^2}{3 \times 10^{-7}} \\ &= 26 \text{ hours.} \end{aligned}$$

At a temperature of 150°C the drying time necessary may be calculated to be 1.6 hours.

Considerable care need be taken both when drying the laminate and when curing the adhesive to avoid damage to the base laminate. Exposure of cured laminates to prolonged high temperature has been reported to cause microcracking in the matrix [31, 32, 34, 37, 38]. Liquid water present in voids or cracks in the laminate may cause delamination and blister formation [34], and liquid water in base honeycomb can cause major core-skin disbonds [39]. It has been suggested that the critical temperature causing matrix cracking is the glass transition temperature of the matrix [38, 40]. The critical factor appears to be the rate at which the temperature is changed so that slow heating and cooling rates may reduce the cracking, delamination, and disbond problems by allowing diffusion to occur without generation of local high pressures due to volatilisation of liquid water in voids, cracks etc. Diffusion, however, is as observed above, extremely slow in structural composites.

Of considerable interest is the observation by Augl [6] that application of vacuum has an insignificant effect on the rate at which moisture is lost from the composite. This is due to the fact that diffusion is the rate determining step in the process and this is affected only in a minor way by small changes in the partial pressure of moisture in the exterior environment which will normally be very low during all drying conditions.

7.7 Design of bonded repairs

Repair concepts have been described earlier. A useful approach for the design of bonded repairs for composite components has recently been described by Baker [41]. This approach is described in the following narrative with the permission of the author.

7.7.1 General considerations

For simplicity, repair designs can be based on a simulated joint configuration obtained by considering a section taken through the damaged region. A more rigorous analysis would need to consider the effective stiffness of the patched region and its influence on the surrounding structure; this, however, is beyond the scope of the present discussion. The patch is usually designed to match the strength and stiffness of the parent material. The strength of the repair can be designed to exceed or equal some level of strain in the composite. This point is usually taken as "B" allowable residual ultimate strain, i.e., that value which is exceeded by 90% of the population with a confidence of 95 percent. One approach would be a design which exceeded the "B" allowable ultimate strain of the undamaged parent material by 20–50%. Less stringent would be a design which restores the reduced "B" allowable ultimate strain of the component in the presence of a stress concentration such as a hole or representative damage. The former approach is the more desirable since it is more conservative and allows for a reserve in strength to accommodate disbonds or defects. For practical considerations it is usually satisfactory to design to the reduced strain levels. For example the "B" allowable for a 16 ply quasi-isotropic laminate with 1% absorbed moisture and statically tested at 120°C after appropriate fatigue loading, is approximately 9000 microstrain for tension and 7500 microstrain for compression. If a strain concentration factor for holes is considered, then the respective strain values are reduced by a factor of three. For practical purposes current designs are based on a "B" allowable of 4000 microstrain for ultimate strain of laminates containing holes or impact damage.

7.7.2 External patch design

An external patch repair can be analyzed as half of a double lap joint as long as the sub-structure provides enough support to minimize bending effects. The behavior of the bondline adhesive must be known under typical service conditions. There are some general assumptions that need to be made regarding a patch repair. First, the patch must have tapered edges in order to minimize peel stresses, particularly when the patch thickness exceeds 8 plies. The stiffness and thermal expansion characteristics of the patch must match that of the parent material. Also, the hole or damage is straight sided, which is a penalty since tapering improves load carrying capability. Then the maximum load carrying capacity of the joint, based on an idealization of the stress strain behavior of the adhesive, is given by

$$p = 2\{\eta\tau_p(\tfrac{1}{2}\gamma_e + \gamma_p)Et\}^{1/2} \tag{7}$$

where τ_p is the effective yield stress of the adhesive, γ_e and γ_p are respectively the elastic strain to yield and the plastic strain to failure, η the adhesive thickness, t the thickness of the patch (and the parent material) and E is the modulus.

As an example, taking (for hot/wet conditions)

τ_p = 20 MPa
E = 72 GPa (typical for graphite/epoxy laminates employed in aircraft)
t = 1.5 mm (12 plies)
γ_e = 0.05
γ_p = 0.5
η = 0.125 mm

gives P = 0.75 kN/mm.

The allowable load (per unit width) in the patch or parent material, P_u, is given by

$$P_u = Ee_u t \tag{8}$$

where e_u is the allowable ultimate failure strain of the composite; taking this as 4000 microstrain gives

$$P_u = 0.43\ \text{kN/mm}.$$

Thus, for the chosen ply thickness in this example, the load capacity of the bonded joint appears to be well above the allowable for the parent material. However, a safe margin is not obtained for external patch repairs for laminates above about 16 plies thick.

If it is considered that the strength of the repair is adequate, the next step is to determine the overlap length. The total patch length is twice this plus the diameter of the hole. The minimum design overlap length (excluding the length of the taper) is given by

$$l = \left(\frac{P_u}{\tau_p} + \frac{2}{\lambda}\right) \times \text{safety factor} \tag{9}$$

where λ (the elastic strain exponent) is given by

$$\lambda = \left(\frac{2G}{\eta E t}\right)^{1/2}, \text{ and where } G, \text{ adhesive shear modulus, is given by } \tau_p/\gamma_e$$

For the present example, using a safety factor of two to provide tolerance to damage such as voids and minor disbonds, gives l as about 60 mm. Thus assuming a hole diameter of 50 mm the total patch length would be 170 mm.

7.7.3 Flush repairs – single scarf configuration

Simple Analysis:

If the patch matches the parent material in stiffness and expansion coefficient, simple theory gives

$$\tau = \frac{P \sin 2\theta}{2t} \tag{10}$$

$$\sigma = \frac{P \sin^2 \theta}{t} \tag{11}$$

where τ and σ are the shear stress and normal stress acting on the adhesive and θ is the scarf angle. At small θ, the normal stress σ is negligible. The required minimum value of scarf angle θ, for an applied load P, can be obtained from the following, taking τ_p as the peak shear stress,

$$P = E e_u t = 2\tau_p t / \sin 2\theta. \tag{12}$$

Thus, for small scarf angles the condition for reaching the allowable strain e_u in the adherends is:

$$\theta < \tau_p / E e_u \text{ radians.}$$

Taking e_u as 4000 microstrain, τ_p as 20 MPa, and E as 72 GPa, gives

$$\theta \leqslant 3^\circ.$$

Thus, the minimum length of the scarf taking the laminate thickness as 13 mm is about 250 mm, which, for a hole size of 100 mm, gives a total patch length of 600 mm.

Laminate repair

The above cited simplified theories assume that the adhesive shear stress is constant. In a composite laminate or patch the stiffness varies with thickness therefore the adhesive shear stress cannot be assumed to be constant. When a composite patch is fabricated the edges are stepped rather than tapered or scarfed but since the steps are quite short the shear stress can be taken to be constant. Such a repair can be considered to be a single lap joint with very short steps.

The distribution of shear stress in the adhesive can be approximately obtained from the following simple analysis which ignores shear lag effects and yielding in the adhesive.

From load equilibrium on each ply step of length Δx, it follows that

$$\tau = \Delta P/\Delta x. \tag{13}$$

If it is assumed that the load increment ΔP on each step is proportional to the relative stiffness of the ply, such that for any layer

$$\frac{\Delta P}{P} \cong \frac{\text{stiffness of ply}}{\text{Total stiffness}} \tag{14}$$

then τ varies through the laminate thickness approximately as does the ply stiffness. Thus, for a $[(0°/\pm 45°/90°)_2]_s$ laminate the stiffnesses are in the ratio

$$1(0°) : 0.23(\pm 45°) : 0.07(90°).$$

This shows (on the basis of the above reasoning) that very high shear stresses occur on the 0° plies. These can lead to shear, or more likely, peel failure, unless they are relieved, for instance by serrating the ends of the plies. Serrations reduce the effective stiffness of the end of the ply and thus act in a similar way to tapering in an external patch.

An approximate estimate of the required step length Δx can be made by assuming the adhesive is stressed to its shear yield stress, τ_p, and each of the plies are loaded individually; then

$$\Delta x = \frac{E_p e_u}{\tau_p} \times \text{ply thickness} \tag{15}$$

where E_p is the ply stiffness (typically around 120 GPa for the 0° plies). Taking e_u as 4000 microstrain, gives a ply length about 3 mm. The length for the $\pm 45°$ and 90° plies could be shorter but in practice would probably be made the same length. Thus, for a laminate 100 plies (13 mm) thick the scarf length is 300 mm.

7.8 Composite service damage experience

The earliest application of composites in aircraft was the use of fiberglass reinforced polyester or epoxy structures. These materials were used in radomes, antenna domes, and secondary structures. The types of structures included both simple laminate and laminate faced honeycomb construction. These components, particularly radomes, projected outwards and were subject to a variety of damage types including erosion, disbonding, and penetration. In this case, repair was considered in the context of restoring electrical integrity to the component since structural loading was relatively light. With repair of such components, bonded repair approaches were used. Monolithic skin components were repaired with external bonded patches. Fiberglass faced honeycomb core construction was repaired by core replacement and an external patch. Of prime importance in such repairs was restoration of the electrical transmission characteristics of the components. Thickness variations, for instance, which changed the electrical characteristics could constitute the basis for rejection of the repair.

A similar repair philosophy was used for structural fiberglass composites. Restoration of some measure of mechanical strength was desired, though not critical since these components were not highly loaded. However, since these components were subjected to some mechanical loading, the repair patches were usually thicker and larger than those placed on radomes. Some of the resins used for the repairs were ambient curing systems which also required no additional pressure for compaction. Some components were exposed to elevated temperatures and the ambient temperature curing resins were not adequate for the task. In these cases, resistance heating blankets were used to provide the elevated cure temperatures required. Additional compaction was provided by vacuum bagging the repair area.

Additional types of reinforcement were subsequently utilized in design of aircraft components. Boron/epoxy (b/ep) components were designed in the late 1960's and early 1970's. There was a marked difference between the capabilities of such a fiber and fiberglass. Primary structure was designed taking advantage of the high modulus and strength of the Boron fiber. The repair of such structure had to be approached in a more scientific fashion. Again, there are various classifications of damage ranging from superficial to extensive substructure damage. For cosmetic purposes, the area need only be filled in with a suitable (temperature) two-part base. Through-the-skin-damage was pursued by two approaches. One approach used a patch to take the load over the damaged area; the other used a scarf type repair. In repairing b/ep components, field repair limitations on damage size were found. Here a patch approach was used to effect the repair. Pretreated Ti foil was used in a stacked arrangement with suitable numbers of layers of film adhesive. Prepreg was also used with the adhesive in a cocuring operation. At times, precured patches could be used in the repair procedure. Autoclaves were used for more involved procedures or larger damage sizes. Where facilities were available, scarf type repairs could be undertaken. Here the area had to be carefully machined, usually with a router. Although Boron composites were used in some military applications, the relatively high cost of the fiber prevented more extensive use. Nevertheless it is used today in hybrid designs or other applications where high rigidity is required along with high strength.

7.9 Specific component repair

AH-1 composite main rotor blade

Repair of composite components has been a concern for rotary wing as well as

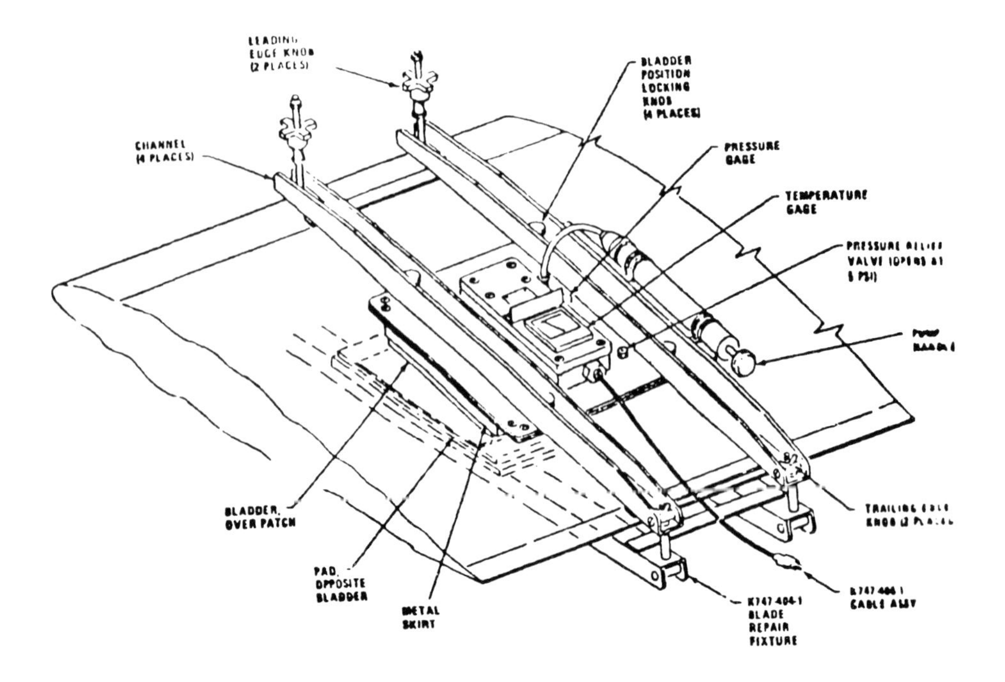

Figure 7.9. Pressure/heat blanket used to repair AH-1 composite rotor blade.

fixed wing aircraft. An effort [42] was undertaken to establish a repair procedure for the AH-1 composite main rotor blade. This particular blade consists of a filament wound S-glass/epoxy spar, a filament wound Kevlar/epoxy basketweave skin and polyimide honeycomb core afterbody. Three types of repair schemes were used for the afterbody structure. One was a skin patch for repair of small punctures and cuts, another was a plug patch for repair of skin and core damage on one side of the blade and a double plug patch for through hole damage. The last method was a V-shaped double patch for repair of the trailing edge. The skin patch developed consists of a precured 3 ply glass/epoxy laminate. The outer ply was 181 fabric with the middle ply being a ± 45 bias double ply with unidirectional E glass and the inner ply of 120 fabric. Damages up to 180 mm in diameter were repaired with these patches. The patches were bonded over the damaged area.

The repair kit designed for these repairs is a pressure/heat blanket as shown in Figure 7.9. Pressure of 30 KPa is supplied by a hand pump, and a rheostat controls the heat blanket to 70–80°C. Cure cycles consisted of 15–30 minutes combined heat and pressure and in some cases an additional 30 minutes of pressure alone. Other stipulations were that the maximum adhesive temperature would be 80°C, that the adhesive be amenable to easy mixing and also that it be spreadable. The adhesive had to be thixotropic to facilitate application to the lower surface of the blade. The adhesives chosen after a screening study were Hysol's EA9330 and Hexcel's HP341.

Fatigue tests were performed on repaired root-end and outboard blade sections.

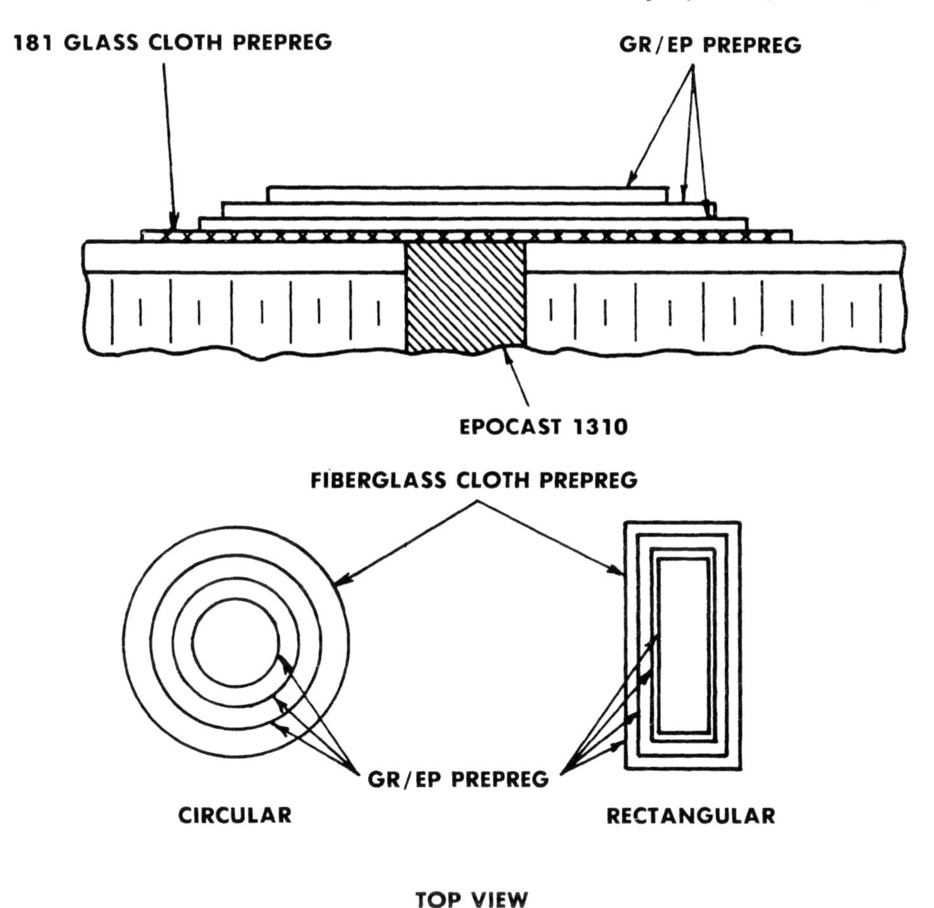

Figure 7.10. Patch configuration for S-3A composite spoiler repair.

All repairs on the blade section showed no degradation or propagation of the original damage. Whirl tests were conducted as a final part of the repair substantiation. A 50-hour whirl test was performed on repaired blades. No deviation in blade characteristics were detected, and the repairs showed no tendency to separate or become unbonded.

These repairs were demonstrated on appropriate rotor blades under field conditions. Blades were flown after each repair for a short 10 to 15 minute flight to confirm that the blades had been restored to acceptable status. The field demonstrations indicated that personnel could complete field repairs in not more than 3 hours.

Repair of S-3A spoilers

A small number of graphite/epoxy spoilers were fabricated from NARMCO 5309/T300 as part of a development program for the S-3A aircraft [43]. One issue was the repairability of such components. There were some general guidelines in

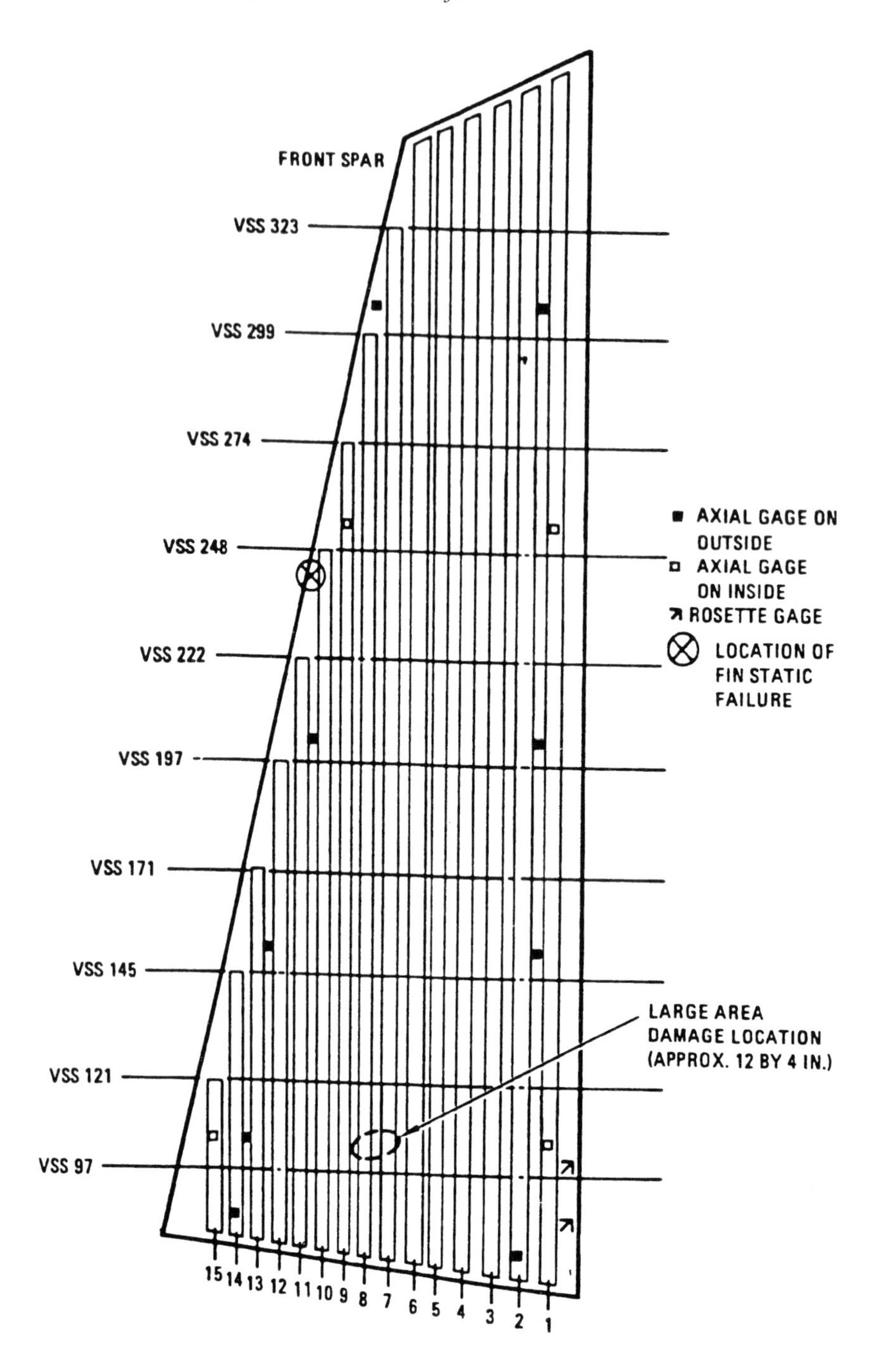

Figure 7.11. Location of repair in damaged L1011 fin.

this program in that the repair should be made using available materials and be cost effective. It was decided that a bonded patch concept would be used, and that the patch would be comprised of matching prepreg material. The patch adhesives were to be 120°C curing systems and the EA product (modulus and cross sectional area) of the patch was required to be twice as great as the damaged skin. The diameter of the repair patch was sized so that the average adhesive shear stress would not exceed 5.9 MPa. The configuration chosen for the patch was a layer of 181 cloth prepreg (used for load introduction from the skin to patch) followed by layers of graphite/epoxy prepreg decreasing in size upward as shown in Figure 7.10. The rectangular configuration was used for scratch damage and the circular for all others. Since the substructure of the spoiler was nomex honeycomb, a core filling epoxy was used for filling in the region previously occupied by damaged core. A 120°C curing film adhesive was used to bond the patch to the spoiler surface.

The spoiler was originally separated into different zones. The zonal location of the damage determined the patch orientation and thickness. The edge configuration did not contain honeycomb and the repair procedure was different for this region. The curing process was accomplished through the use of vacuum heating blankets. These repaired spoilers have been flying on S3A aircraft and have met all performance requirements.

Repair of the vertical fin of the L-1011

Repairs have also been conducted on commercial aircraft such as the L-1011. In this case, a vertical fin from an L-1011 was selected as the item for study [44]. Figure 7.11 shows a damage 300 mm 100 mm simulating lightning strike damage on the vertical fin. This damage was simulated by impacting to obtain delamination and by burning through the skin with a welding rod. The area was then charred by heating with an oxygen acetylene flame torch.

The repair concept used for this effort was based on a precured bonded external graphite/epoxy patch. Figure 7.11 depicts the repair location on the damaged fin. The patch was sized to match the EA product of the skin and the stiffener flanges. No 90° plies were included to avoid a large difference in Poisson's ratio between the patch and fin cover. The length of the patch was sized so that the maximum average shear stress did not exceed 3.5 MPa. This conservative approach was based on the possibility of moisture in the fin laminate and potential voiding in the bondline. The repair was sized to carry 150% of design ultimate load for the fin cover to ensure it would not fail during residual strength testing. As Figure 7.12 shows, precured 4-ply (45, −45, −45, 45) and 3-ply O_3 patches were bonded in sequence for the repair. The 4-ply patch was fabricated from Narmco T300/5208 and the 3-ply patch from AS4/3502 prepreg material. All patches were ultrasonically inspected and found to be void free. Before the application of the patch it was necessary to clean out the burned through hole. A precured graphite/epoxy disc was then bonded into the hole with EA9330 two-part room-temperature curing adhesive system. A layer of Narmco Metalbond 329 adhesive was then applied to each surface of the five precured layers and the layers laid up maintaining a 6 mm step at each layer. The repair was vacuum bagged and cured at 177°C for 2 hours.

The cured patch was inspected with an ultrasonic technique and bondline voiding was detected. Unaugmented vacuum cure was given as the primary reason for the presence of bondline voids. The hat stiffeners were repaired by mechanically attaching hat flanges to the repaired skin after the graphite/epoxy patch was inspected. After the completion of the repair, the vertical fin was subjected to one lifetime of fatigue cycling using the appropriate spectra. The repaired article was subsequently loaded statically until failure. There were no apparent effects of the fatigue loading and the article failed statically at 120% of design ultimate load with failure occurring well away from the repair area.

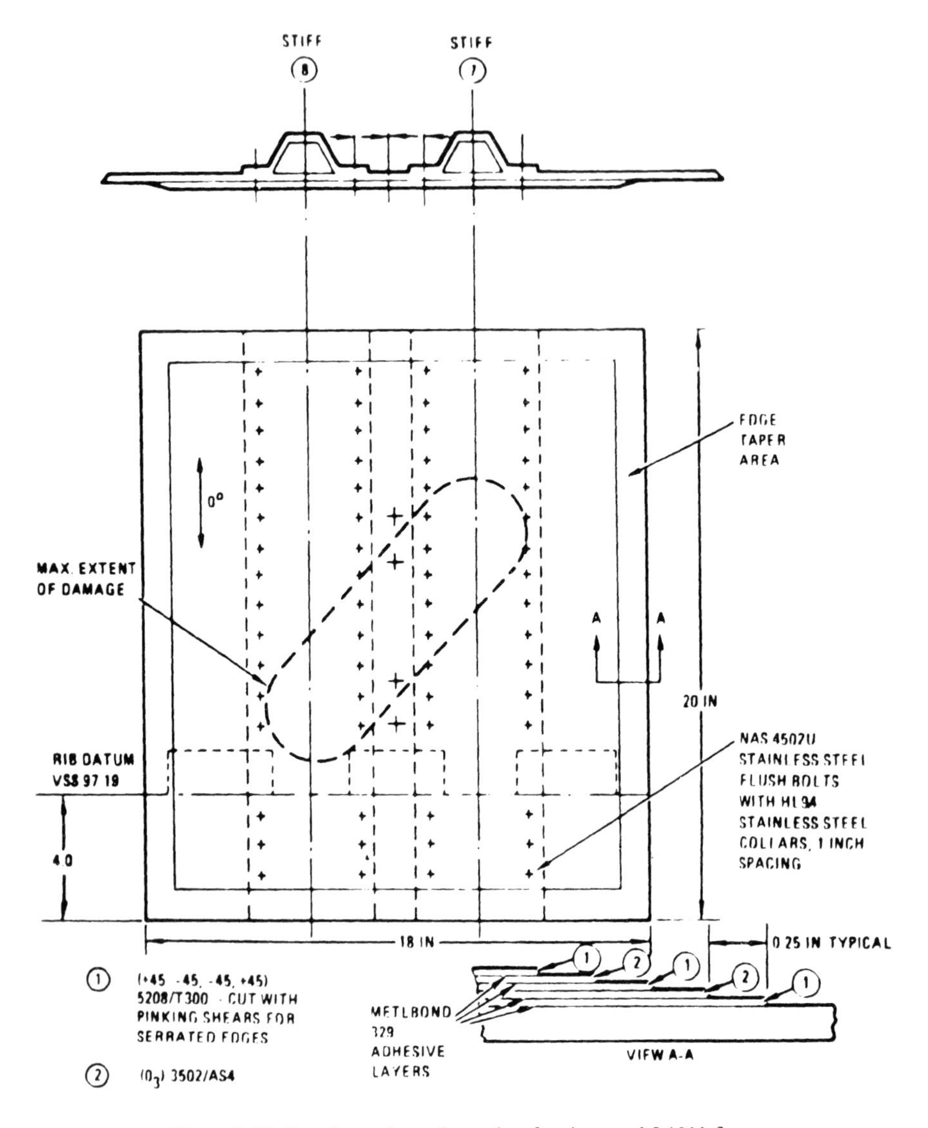

Figure 7.12. Repair patch configuration for damaged L1011 fin.

7.10 Future requirements

Bonded repair offers a viable approach to repair damaged composite components. Properly designed and processed bonded repairs can restore more of the load carrying capability of a component than a bolted approach. Nevertheless, there are specific areas which must be addressed to successfully implant this methodology as part of the maintenance philosophy. The problem of moisture and subsequent bondline voiding is a continuing concern. Complete drying prior to repair is not a practicable approach to the problem. Alternative methods must be used to obviate this problem. New materials and processes are required to simplify the repair procedure particularly for the remote or field situation. The requirement for elevated temperature and pressure cures must be minimized as much as possible while storage life must be maximized.

Along with implementation of new technology is the requirement for adequate training and preparation of personnel. Proper instruction and preparation of responsible individuals is necessary for assigned repair tasks. Dissemination of the appropriate instruction manuals on the subject is also required.

References

[1] W.W. Stinchcomb, "Technology of Composite Materials," ASTM Standardized News, pp. 12–15, December, (1983).

[2] James M. Whitney, "Composites in the Aircraft Industry," ibid pp. 16–19.

[3] David Velupillai, "Carbon into Airline Service," Flight International, pp. 364–000 14 August (1983).

[4] R. Weinberger, A.R. Somoroff and B.L. Riley, "U.S. Navy Certification of Composite Wings for the F-18 and Advanced Harrier Aircraft," AIAA/ASME 18th Structures, Structural Dynamics, and Materials Conference pp. 396–407, March 24–26, (1977).

[5] George Lubin, "Handbook of Fiberglass and Advanced Plastics Composites," Van Nostrand Reinhold Company, New York, (1969).

[6] Clayton A. May, Editor, "Resins in Aerospace," ACS Symposia Series No. 132, American Chemical Society, Washington, (1980).

[7] J.E. Ashton, J.C. Halpin and P.H. Petit, "Primer on Composite Materials Analysis," Technomic Publishing Co., Connecticut, (1967).

[8] Robert M. Jones, "Mechanics of Composite Materials," McGraw Hill Book Company, New York, (1975).

[9] Stephen W. Tsai, "Introduction to Composite Materials," Technomic Publishing Company, Connecticut, (1980).

[10] "Carbon and Graphite Fibers; Manufacture and Applications," M. Sittig, Ed., Noyes Data Corp., p. 356, (1980).

[11] R.B. Pipes, Editor, "Nondestructive Evaluation and Flaw Criticality for Composite Materials," ASTM STP 696, American Society for Testing and Materials, Pennsylvania, (1979).

[12] "Characterization, Analysis, and Significance of Defects in Composite Materials" Proceedings of the 56th Meeting of the AGARD Structures and Materials Panel, London, England 10 to 15 April (1983).

[13] S.N. Chatteigee, R.B. Pipes and B. Dick, "Composite Defect Significance," Annual Tech. Report MSC TFR 1409/1109, Materials Science Corporation.

[14] J.C. Duke, Jr., "Nondestructive Evaluation of Composite Materials: A Philosophy, An Approach, and an Example." pp. 75–95, ASTM STP 797, "Composite Materials: Quality Assurance and Processing," Editor, C.E. Browning, American Society of Testing and Materials, Pennsylvania, (1981).

[15] I.M. Daniel and T. Liber, "Nondestructive Evaluation Techniques for Composite Materials," Symposium on Nondestructive Evaluation Proceedings, San Antonio, Texas, pp. 226 April 24–26, (1979).
[16] Radiography, ibid. pp. 232.
[17] M.L. Phelps, "In Service Inspection Methods for Graphite/Epoxy Structures on Commercial Transport Aircraft," NASA CR-165746 of November (1981).
[18] Lubin, G. *et al.*, "Repair Technology for B/Epoxy Composites," Grumman Aerospace Corporation, AFML-TR-71-270, February (1972).
[19] Crabtree, D., "Adhesives for Field Repair of Graphite/Epoxy Composite Structures," NADC Report 79286-60, September (1981).
[20] Beck, C.E. and Myhre, S., "Repair Concepts for Advanced Composite Structures," Journal of Aircraft, Vol. 16, No. 10, p. 720, October (1979).
[21] Deaton J.W. and Mosso, N.A., "Preliminary Evaluation of Large Area Bonding Processes for Repair of Graphite/Polyimide Composites," 28th National Sampe Symposium, 12–14 April (1983).
[22] Delano, C., McLeod, A., Riel, F. and Greer, R., "A New Resin for Field Repair," 13th National Sampe Technical Conference, 13–15 October (1981).
[23] Gillman, H.D. and Eichelberger, J.L., "The Application of Ultraviolet Cure Resins for Repair of Composites," Report on Navy Contract N62269-77-M-7197, November (1977).
[24] Jaquish, J., Sheppard, C. and Hill, S., "Graphite Reinforced Thermoplastic Composites," Report on Navy Contract N00019-79-C-0203.
[25] J.C. Watson, "Bolted Field Repair of Composite Structures," NADC Report No. 77109-30 of 1 March (1979).
[26] R.E. Bohlmann, G.D. Reniere and M. Libeskind, "Bolted Field Repair of Gr/Epoxy Wing Skin Laminates," ASTM Symposium: Joining of Composites, Minn, Mn, 15–16 April (1980).
[27] Horton, R.E. *et al.*, "Adhesive Bonded Aerospace Structures Standardized Repair Handbook," Boeing Commercial Airplane Company, Seattle, AFFDLTR-77-139, October (1977).
[28] C.H. Shen and G.S. Springer, "Moisture Absorbtion and Desorbtion of Composite Materials," *J. Comp. Mat.*, 10, 2, (1976).
[29] A.C. Loos and G.S. Springer, "Moisture Absorbtion of Graphite-Epoxy Laminates Immersed in Liquids and in Humid Air," *J. Comp. Mat.*, 13, 131, (1979).
[30] G.S. Springer, "Moisture Absorbtion in Fiber-Resin Composites," Developments in Reinforced Plastics –2, Ed. G. Pritchard, Elsevier Applied Science.
[31] D. Shirrell, "Diffusion of Water Vapour in Graphite/Epoxy Composites, "Advanced Composite Materials – Environmental Effects," ASTM STP 658, Ed. J.R. Vinson, American Society for Testing and Materials, 1978, pp. 21–42.
[32] J.M. Whitney and C.E. Browning, "Some Anomolies Associated with Moisture Diffusion in Epoxy Matrix Composite Materials," ASTM STP 658, loc. cit., pp. 43–60.
[33] J.M. Augl, "Moisture Transport in Composites During Repair Work," 28th National SAMPE Conf. April 1983, "Materials and Processes – Continuing Innovations," Vol. 28, 273, (1983).
[34] S.H. Myhre, J.D. Labor, and S.C. Aker, "Moisture Problems in Advanced Composite Structural Repair," *Composites* 13, 289, (1982).
[35] T.M. Donnellan, J.G. Williams and R.E. Trabocco, "Void Formation in Adhesive Bonds," SAMPE Preprints 15th National SAMPE Technical Conference (1983).
[36] See for example J.D. Verhoeven, "Fundamentals of Physical Metallurgy," Wiley, New York (1975), pp. 137–146.
[37] H.T. Hahn and R.Y. Kim, "Swelling of Composite Laminates," ASTM STP 658, loc. cit. pp. 98–120.
[38] E.L. McKague Jr., J.E. Halkias, and J.D. Reynolds, "Moisture in Composites: The Effect of Supersonic Service on Diffusion," *J. Comp. Mat.* 9, 2, (1975).
[39] R.A. Garrett, R.E. Bohlmann and E.A. Derby, "Analysis and Testing of Graphite/Epoxy Sandwich Panels Subjected to Internal Pressures Resulting from Absorbed Water," ASTM STP 658, loc. cit. 234–253.
[40] P. Shyprykevich and W. Wolter, "Effects of Extreme Aircraft Storage and Flight Environments on Graphite/Epoxy." ASTM STP 658, loc. cit. pp. 234–253.

[41] B.C. Hoskin and A.A. Baker, "Composite Materials for Aircraft Structures," published in AIAA Education Series, 86, Chapter 6 Repair of Graphite/Epoxy Composites by A.A. Baker.
[42] A.S. Falcon, "Field Repairs for the AH-1 Composite Main Rotor Blade," 11th National SAMPE Conference, November 13–15, (1979).
[43] C. Foreman, S.A. McGovern and R. Knight, "S-3A Gr/Epoxy Spoiler Fabrication of Ten Shipsets and Damage Repair Study," Report No. NADC-76234-30 of May (1976).
[44] Stone, R., "Development, Demonstration, and Verification of Repair Techniques and Processes for Graphite/Epoxy Structures for Commercial Transport Applications," NASA Contractor Report 159056 of January (1983).

Index